THE AI FACTOR WORKBOOK

Also by Asha Saxena

The AI Factor: How to Apply Artificial Intelligence and Use Big Data to Grow Your Business Exponentially

THE AI FACTOR WORKBOOK

A PLAYBOOK FOR ACHIEVING EXPONENTIAL GROWTH

ASHA SAXENA

A POST HILL PRESS BOOK
ISBN: 979-8-88845-693-4

The AI Factor Workbook:
A Playbook for Achieving Exponential Growth

Cover design by Conroy Accord

This is a work of nonfiction. All people, locations, events, and situations are portrayed to the best of the author's memory.

Post Hill Press
New York • Nashville
posthillpress.com

Published in the United States of America
1 2 3 4 5 6 7 8 9 10

To my mom, Raj Khanna, who was my biggest fan,
my cheerleader, and taught me to work hard,
be ambitious, and never give up.
To my dad, Bhim Khanna, the epitome of kindness.
To my husband, Rajeev Saxena, who brings stability
and intellect to our family unit.
To my kids, Siddharth and Milan Saxena,
who are my drivers to always excel.
https://ashasaxena.com

And to my WLDA community, who inspire me
to create a fair digital world with parity and equity.
This community embodies the three pillars
of our shared vision: Community - a peer-to-peer network of
senior leaders in data and AI;
Growth - maintaining a growth mindset
in the ever-changing, fast-evolving tech world;
and Impact - striving to leave the world a better place
by building future leaders in data and AI.

To The AI Factor community, always investing
in learning, growing, and implementing the fast-evolving
changes to stay ahead. Your dedication and commitment to
progress are the driving forces behind continuous innovation
and excellence in the realm of AI.
https://theaifactor.ai

Contents

INTRODUCTION

How to Use This Book

Late in 2022, the launch of ChatGPT and other generative AI products started a widespread media feeding frenzy. Business leaders and ordinary workers alike experienced sudden panic and uncertainty. "What should we do? Should we just start using all the available AI gadgets and see what happens? Should we drop everything and add AI to every business process?" And above all, the big question on everyone's lips was, and still is, "Where do we start?"

Based on three decades of experience in data management, analytics, machine learning, and AI, I have devoted my career to strategy, consulting, mentoring, and speaking on the subject that led to the writing of my first book, *The AI Factor*. It was published in mid-February 2023, and the timing could not have been more interesting.

Since its release, I have received a growing number of requests to hold training sessions, helping business leaders and their key people put the principles of *The AI Factor* into practice. It soon became apparent that we also needed a systematic *playbook* to orient them and their teams of AI's best practices. We also needed a project-centered *workbook* to help them map out and execute specific plans for AI implementation. The result is the work you are now reading.

Each of the sections of this book is divided into two parts. The first is the playbook portion—mapping out basic principles and inviting group discussion on the strategic and tactical requirements of a sound AI implementation. That is followed by the workbook. As you go through the workbook section, start to write down your own outline—and the first of many plans to add sound AI strategy to your organization.

Methodology Note

When planning your AI strategy, the first step is to consider *an expanded, high-level view* of the subject at hand. Consider as many potential AI strategies or projects as your type of business could conceivably benefit from. In the workbook, you will have ample space to record these ideas for the future. However, the second step is take *a focused, detailed view.* Many different projects may occur to you as you go through this process, but be sure to focus first on a single AI project. It will ultimately lead to others, but it is important to know intimately the elements that lead to successful, practical AI.

Learning Objectives

This work's purpose is to equip you with the practical knowledge required to design and implement artificial intelligence and data strategies that will benefit your organization, its members, and its customers. It won't make you a data scientist, but it will inform and help you build and work with a team (including data scientists) in a focused, goal-oriented manner. Following the steps outlined in this workbook will help you understand and develop an AI strategy for your organization by understanding the basic elements of artificial intelligence.

Each of the book's sections is designed to define and meet specific learning objectives. These can be applied and measured informally, by individuals going through the material on their own. If the book is being used as part of a formal training program, then these objectives may be applied, and performance measured accordingly. In brief, these objectives are:

Analyzing Business Quadrants: The first objective is to have a clear, practical understanding of the type of organization you belong to when it comes to using AI. This classification is based on the optimizer-extender-innovator-multiplier model set forth in *The AI Factor*. Different types of companies have different business priorities, and so the ideal first AI project for your company must be tailored to meet those priorities. The workbook section will help you determine your own organization's type.

Mastering Data Readiness: You will assess the current state of the organization's existing data, including whether or not the data is siloed or integrated, structured or unstructured, owned by the organization, and adequate for use in AI. The workbook section will cover your organization's *general* data readiness, but it will focus mainly on data readiness issues that will relate to a single project.

Sustainable, Responsible, and Ethical AI: The objective here is to know what makes AI a technology to be embraced rather than feared or resisted. It will cover issues involving data privacy, bias, transparency, explainability, and adherence to human values. In the workbook, following the responses to general sustainability questions, you will be asked to provide specific, focused answers to sustainability issues related to your project defined in the previous section.

Finding and Aligning the Right People: The objective here is to identify, in general, the type and qualifications of individuals necessary to a successful AI implementation. This will include the "buy-in" requirements from key individuals that indicate an organization's maturity and internal competence. In the workbook section, you will need to be specific about job requirements for the target project.

Prioritization Techniques: The next objective is to understand what types of AI and data projects have the best combination of data quality and business value. In the workbook section, you will develop a hypothetical, high-priority AI project based on your organization's type and data readiness. The goal is to develop a focused view and strategy for a high-priority AI initiative for your organization through the exercises in this workbook.

Understanding AI Implementation: The playbook part of this section will review the general requirements of identifying and implementing AI and data projects. This will include principles explained in previous sections, but it will also go into detail on how to define the parameters of data-centric projects and how to establish measurable project goals. In the workbook section, you will be asked to describe those steps as they pertain to the target project.

Scalability (The AI Flywheel): The learning objectives involve how to plan, prototype, test, measure, and ultimately scale your AI project. The workbook will test your knowledge of the general principles, but it will focus mainly on filling in production details for the target project. This will include a framework for measuring success.

A Practical Exercise

When using this book, consider leveraging any of the wide array of available AI tools to help with some initial idea generation and brainstorming. These include tools like ChatGPT, Perplexity, Bard, Gemini, and likely new ones that will emerge shortly. Do this at any point in the workbook sections as a valuable assist in the learning process.

Some example prompts could be any of the following:

> **"For a company in the [insert your industry] industry, where could AI drive efficiencies and increase revenue?"**
>
> **"For [insert your company name], where could AI drive efficiencies and increase revenue?"**
>
> **"What is the typical process flow for a B2B [insert your type of company] company?"**
>
> **"How could AI improve this process flow?"**
>
> **"For a software as a service company, what are the compliance, legal, and ethical considerations I need to be aware of?"**
>
> **"What are the business value drivers of enhancing call center effectiveness?"**

Of course, you can customize and develop your own prompts. In that way, you'll be leveraging AI tools to advance your understanding of AI's potential for your organization.

The Opportunity Ahead

Artificial intelligence and the use of large data sets are not deep, incomprehensible mysteries. They are neither dire threats nor get-rich-quick schemes. AI is a tool, albeit a powerful one, that can help companies, nonprofits, and the people who work in them be more efficient, more productive, and better able to create new approaches and drive innovative ideas. While the technology is advancing at a very rapid pace, the techniques identified in this workbook provide a framework that is adaptable and can be used to align the technical approach toward business value. By following this workbook's practical steps, you can realize these benefits and create new opportunities for success.

Let's get started.

Asha Saxena

PART I

ARE YOU READY FOR AI?

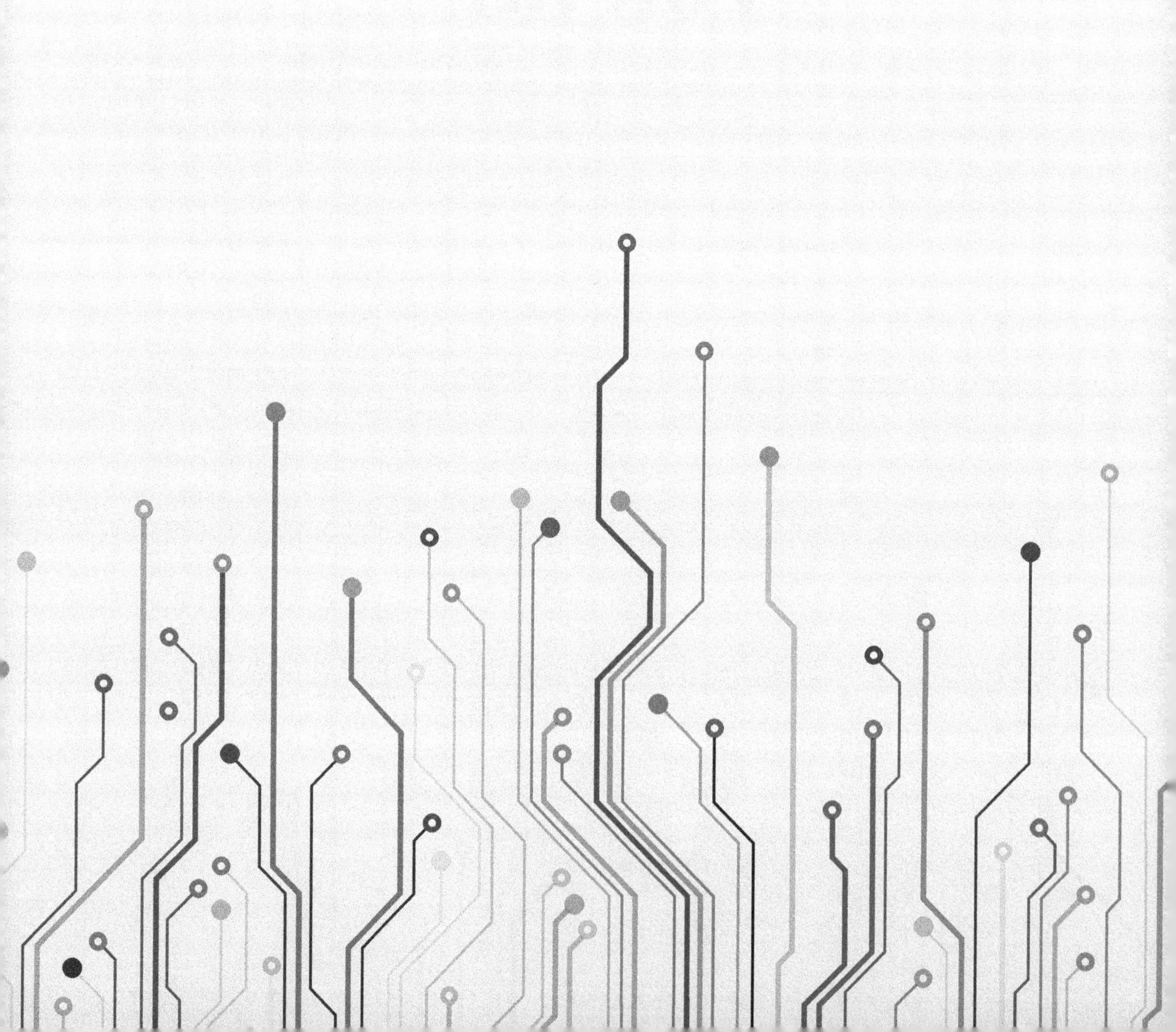

SECTION ONE

Analyzing Business Quadrants

Before you start on your journey with AI, large or small, you must first identify where you are in your organization today, starting with understanding your business goals, your organization's overall strategy and focus, its capabilities, resources, and infrastructure/platform that form the basis of a data-driven AI strategy. By doing so, you will understand the current stage and innovative potential of your business. Once that is known, you will be better able to identify the AI projects that could prove most beneficial. Ultimately, this will lead to scaling the project, developing similar AI and data projects, and even transforming the organization itself.

PLAYBOOK

In *The AI Factor,* I described two important factors that will help you understand the type of organization to which you primarily belong. Most companies do not fit neatly into only one category, of course, but the two factors are:

- *Potential for growth*—This depends on many factors, such as market size and the number and nature of your competitors.
- *Willingness to innovate and take risks*—This is more difficult to measure. Management style and a company's track record are good indicators here.

With these factors in mind, organizations can be characterized in four ways. Determining what type they are, in general, will indicate what kinds of AI and data-intensive projects will likely help them succeed.

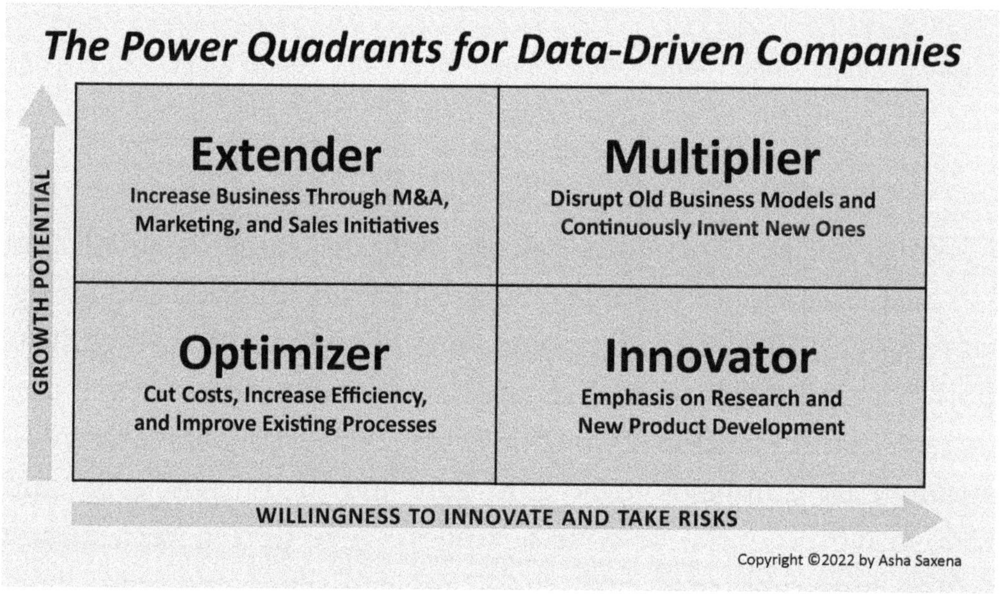

The Power Quadrants for Data-Driven Companies

GROWTH POTENTIAL	**Extender** Increase Business Through M&A, Marketing, and Sales Initiatives	**Multiplier** Disrupt Old Business Models and Continuously Invent New Ones
	Optimizer Cut Costs, Increase Efficiency, and Improve Existing Processes	**Innovator** Emphasis on Research and New Product Development

WILLINGNESS TO INNOVATE AND TAKE RISKS

These four types are not static or immoveable. Companies that fall under the Optimizer category, for example, can migrate to becoming Innovators or Extenders, especially if they employ AI and data solutions in meaningful ways. Also, it is important to note that some of the sample AI projects described below may apply to more than one type of organization.

Characteristics of an Optimizer Organization

Companies in this quadrant can be of any size or age but the term is often associated with relatively large, well-established firms. They can also be functional areas within a company, such as the financial, supply chain management, or sales and marketing departments, especially where the focus is on optimization strategy. Inherently cautious and conservative by nature, such companies tend to focus on bottom-line cost savings, optimizing the process efficiency, reducing potential liability, and protecting their intellectual property. These are all good business practices, but they may indicate a company's aversion to risk and change or a limited growth potential.

Optimizers can nevertheless benefit from the use of AI and data-centric initiatives. One reason for this is that large, well-established companies typically already have an abundance of operational and transactional data. This includes data on customer preferences, purchase behavior, and the company's service history. Once those forms of data are validated, AI models may be designed and successfully implemented.

Some examples of AI projects suitable for Optimizer organizations include the following (not an exclusive list):

Description	Data Sources	AI Technique(s)	Benefits (and Risks)
Identify and reduce inefficiency in an internal business process, such as product development, sales processes, and financial planning and analysis.	Routinely status reports and HR data stored in a company's internal databases. Sales, marketing, and financial data stored in corporate CRM, ERP, and other systems.	• Process mining • Process automation • Machine learning and deep learning • Robotic process automation • AI model optimization	Greater management efficiency and faster progress towards business goals. (Potential workforce reduction, cost savings with negative impact on morale.)

Shorten customer service call times and increase customer satisfaction.[1]	Call center analytics, such as customer service recordings and call logs, survey data, and service history. Text analytics from live chat, SMS, and mobile apps. Speech analytics data from customer interactions.	• Intelligent self-service chatbots and voice bots • AI-based agent coaching • Conversational AI-based agent assistance • Post-call analytics to give insights and summarize the call.	Reduction in human error and increase in accuracy, improved customer service and satisfaction. 24/7 availability, leading to faster response times and increases customer satisfaction. Automation of repetitive tasks, resulting in improved efficiency and productivity.
More efficient handling of sales- and support-related queries.	Customer data, including customer characteristics, preferences, and purchase history. Prospect, market, sales performance, and product information data.	• CRM integration and sales enablement • Pipeline management • Automated ticketing system • Conversation analysis • Task automation • Competitor review assessment	Calls automatically qualified, classified, and correctly routed, resulting in improved customer experience providing personalized and prompt responses. (Potential workforce reduction, and risk of customer alienation.)
Identify more profitable products or services to be used in marketing campaigns.[2] (Applicable to Extender organizations.)	Product and inventory data, product purchase history, and customer reviews. Sales history and finance data.	• Predictive analytics • Personalized product recommendations • Targeting tools • Data analytics for marketing campaigns • Performance optimization	More accurate and real-time targeted campaign optimization. Enhanced personalization and targeting resulting in increased revenue. (Very little risk.)

1 Bamberger, Simon, et al. "How Generative AI Is Already Transforming Customer Service." BCG Global, 6 July 2023, www.bcg.com/publications/2023/how-generative-ai-transforms-customer-service.

2 Briggs, Fiona. "Retail Gets Personal in 2024, Says Akeneo." Retail Times, 16 Jan. 2024, https://retailtimes.co.uk/retail-gets-personal-in-2024-says-akeneo/.

Reduce the cost of recruiting, hiring, and onboarding the right employees.[3] (Such a project is also applicable to Extender, Innovator and Multiplier organizations.)	Resumes, letters, social media profiles, application form data. Recruitment and employee referral data. Cost per hire and onboarding costs data.	• AI recruiting tools for Resume screening and candidate sourcing. • Automated candidate sourcing • Predicting candidate success and culture fit • Onboarding automation and personalized onboarding • Applicant tracking systems	Faster, more efficient HR process. Cost reduction, greater efficiency and speed, Improved candidate matching, personalized candidate experience, enhanced onboarding for new hires. (Potential HR workforce reduction, see above, and risks of built-in bias, skewing or gaming the system.)
Predict routine maintenance or replacement tasks for company-owned or leased equipment.[4]	Device sensor data. Maintenance history records and data on the age, date of purchase, environmental data, and current condition of the equipment.	• Advanced modeling • Anomaly detection • Real-time monitoring and anomaly detection • Machine learning models to analyze large volume of data. • Vibration analysis	Less productivity loss from unexpected equipment failure. Lower cost of lease service agreements. Increased equipment lifespan, optimized maintenance processes, improved safety, and minimized downtime. (Very little risk.)

Characteristics of an Extender Organization

In markets where potential growth is obvious, but the company is wary of risk or change, the extender strategy is very common. Tried and true manufacturing, product development, or service

3 Kelly, Jack. "How Companies Are Hiring and Reportedly Firing With AI." Forbes, 7 Nov. 2023, www.forbes.com/sites/jackkelly/2023/11/04/how-companies-are-hiring-and-firing-with-ai/.

4 "First Step in AI-Based Predictive Maintenance." Sensemore, 9 Jan. 2024, https://sensemore.io/first-step-in-ai-based-predictive-maintenance/.

methods are in place, and the company has achieved success in its existing sphere. What distinguishes extender from optimizer companies is an overwhelming desire or mandate to grow—to conquer new worlds, so long as the new worlds look much like the current ones.

Extenders tend to grow in two major ways, both of which can greatly benefit from the use of data and AI. They can grow through mergers and acquisitions (M&A), by improving its sales and marketing process, or by upselling, reselling, or launching new products.

It's important to note that Extenders *may* find value in AI projects more commonly associated with Optimizer organizations. After all, finding ways to reduce costs or otherwise increase efficiency are common to every business. However, given its focus on growth, an Extender would usually be better served, at least initially, by AI projects such as the following examples:

Description	Data Sources	AI Technique(s)	Benefits (and Risks)
Identify companies with similar products or services that would be likely candidates for profitable acquisition.[5]	Public financial records, media reporting, social media, results of due diligence efforts.	• Target identification • Due diligence automation • Integration prediction • Valuation models • Regulatory and compliance checking • Risk assessment • Stakeholder sentiment analysis	More reliable, advance knowledge of a candidate's financial health and reputation. Reduced risk of making a poor M&A decision or overpaying for a company. (The risk is that an AI approach may miss critical aspects of a company's health.)

5 Cremades, Alejandro. "How to Leverage AI Technology to Win Your Next M&A Deal." Alejandro Cremades, https://alejandrocremades.com/how-to-leverage-ai-technology-to-win-your-next-ma-deal/. Accessed 1 Feb. 2024.

Identify and acquire new customers currently buying similar products from rival companies.[6]	CRM data, web analytics, market segmentation, surveys and other market research and media reporting, social media data, search data, sensor and IoT data.	• Personalized interactions • AI-enhanced marketing content • Behavior pattern analysis • Advanced customer segmentation	Creation of a list of qualified prospects for the marketing and sales teams. (Very little risk.)
Develop new ways to address customer requests.[7]	Internal communication data, web analytics, recorded customer conversations, surveys and other research, publicly available unstructured data.	• Support ticket automation • Improved customer personalization • Customer insight analysis	Greater overall customer satisfaction. Better insights into customer concerns.
Create a more effective automated process for outbound emails, calls, or other queries to qualified prospects.[8]	CRM data (including purchase history), survey and research results, social media profiles.	• Repetitive sales task automation • Customer data analysis • Machine learning for sales engagement • Improved sales planning and reporting	Improved response rates and volume of closed business. (Risks include potential prospect alienation.)
Create an effective, automated process for handling inbound marketing[9] and other web queries.	Product data, CRM data (including purchase history), internal company process data.	• AI chatbot creation and integration • Task and workflow automation • Intelligent call routing	Improved sales cycle efficiency and volume of closed business.

6 Davenport, Thomas H., et al. "How to Design an AI Marketing Strategy." Harvard Business Review, 30 Aug. 2021, https://hbr.org/2021/07/how-to-design-an-ai-marketing-strategy.

7 Heaslip, Emily. "How AI Will Help You Serve Your Customers Better." CO, US Chamber of Commerce, 2 Sept. 2023, www.uschamber.com/co/grow/customers/ai-customer-service.

8 "Sales Automation AI: How Artificial Intelligence Improves Sales." AI Marketing Spot, 24 Oct. 2022, https://aimarketingspot.com/sales-automation-ai/.

9 MTS Staff Writer. "AI's Impact on Inbound Marketing." MarTech Series, 3 Nov. 2023, https://martechseries.com/mts-insights/staff-writers/ais-impact-on-inbound-marketing/.

Create a larger referral network of satisfied customers and "net promoters."[10]	Detailed history of past sales and interactions, social media reviews, survey and research results.	• Personalized service and communications • Predictive routing of inquiries to best-qualified agent • Automated task management	Increased referral-originated business. (Risks include possible alienation and skewed data from social media attacks by competitors.)

Characteristics of Innovator Organizations

The next quadrant—Innovators—includes those aggressive, risk-taking companies whose growth potential is not yet obvious, at least to the outside world. Typical examples of Innovators include tech startups, medical and other science-focused companies, and entrepreneurs of almost every type and size. It also includes individuals and groups within larger companies who innovate within their own department or division. Others include traditional service providers and manufacturers looking to transcend their existing offerings and broaden their customer base. An Innovator's chief characteristics are ambition, vision, exploration of the unknown, and a desire to change things for the better, even before their ideas for products or services are proven successful.

Typical AI projects for Innovator organizations involve research and development for new products or services—or for improvements to existing offerings. This is often coupled with aggressive search efforts to secure the top talent required to create new products and services. These companies invest heavily in market research, new product idea generation, feasibility studies, prototype development, user testing and feedback, technology and material research, and IP management.

The following examples of AI projects for such organizations are typical, but by no means exclusive:

10 Burns, Maureen., et al. "Using AI to Build Stronger Connections with Customers." Harvard Business Review, 1 Aug. 2023, https://hbr.org/2023/08/using-ai-to-build-stronger-connections-with-customers.

Description	Data Sources	AI Techniques(s)	Benefits & Risks
Improve the market research process.[11]	Historical market performance data. consumer behavior data, competitor data, unstructured data (e.g., social media posts).	• Natural language processing (NLP) • Predictive analytics • AI-enabled search and data extraction • Automated survey and report generation	Automate data collection, analysis, and reporting. Increased accuracy of predictive modeling. (Requires precise data. May be subject to bias.)
Develop new products and technology based on expert knowledge, meaningful usage patterns and relevant training data.[12]	Meaningful process descriptions, unstructured image data.	• Pattern recognition and analysis • Deep learning • Artificial neural networks	Business advantage of being the first to develop breakthrough technology.
Improve product feasibility research.	Market and product data, customer feedback, historical and supply chain data, unstructured text and image data.	• Natural language processing (NLP) • Machine learning • AI-powered recommender systems	Better understanding of customer preferences and behavior. Identify new product opportunities based on non-subjective criteria.
Improve the product prototyping and testing process.[13]		• Machine learning • Predictive analytics • AI-driven test simulation and optimization	

11 Hughes, Chase. "Harnessing AI For Market Research: Opportunities And Challenges." Forbes, 5 Oct. 2023, www.forbes.com/sites/forbesbusinesscouncil/2023/08/30/harnessing-ai-for-market-research-opportunities-and-challenges/.

12 Columbus, Louis. "10 Ways AI Is Improving New Product Development." Forbes, 9 Jul. 2020, www.forbes.com/sites/louiscolumbus/2020/07/09/10-ways-ai-is-improving-new-product-development/.

13 Perkowski, Roman. "How Is AI Used in Product Prototyping?" TS2 Space, 16 Sept. 2023, https://ts2.com.pl/en/how-is-ai-used-in-product-prototyping/.

Accelerate software development coding tasks.[14]	Existing code examples and documentation, both internal and open source.	• Faster prototyping, debugging, testing and code analysis • Code explanation • Testing and software performance monitoring	Faster deployment and testing of new or improved software.

Characteristics of Multiplier Organizations

When a company or organization knows its true growth potential and is fully prepared to innovate and take risks to realize that potential, then it will be one willing to bend the rules. This does not mean violating individual rights, or ethical norms, or exploiting others unfairly for financial gain. What I mean by rule-bending is a willingness to challenge the *status quo*. Companies like Netflix, Uber, and Airbnb did so by discarding old models and practices, perceiving shifts in consumer behavior, and building new business models from the ground up.

Becoming a multiplier does not happen overnight, nor can you disguise reckless or destructive business practices under a multiplier label. Once you embrace a mentality of innovation and growth, however, the use of big data and AI will be a perfect fit, enabling you to detect meaningful patterns and act upon them.

Netflix, for example, became an astute observer of customer behavior—notably their dissatisfaction with rental late fees. By creating an online queue process (thus eliminating late fees altogether), it quashed its main competitor, Blockbuster. In the process, it also created a massive set of consumer preference data. Based on AI and machine learning analysis of the data, it modeled consumer preferences and then created new content with a high degree of confidence in its success.

14 Gewirtz, David. "Implementing AI into Software Engineering? Here's Everything You Need to Know." ZDNET, 12 Oct. 2023, www.zdnet.com/article/implementing-ai-into-software-engineering-heres-everything-you-need-to-know/.

Note that Multiplier organizations will often need to deploy AI strategies common to Optimizer, Extender, and especially Multiplier organizations, such as ways to streamline the recruiting and hiring process. However, the types of AI projects that are unique to Multiplier organizations include the following:

Description	Data Sources	AI Technique(s)	Benefits & Risks
Develop a new line of business based on data derived from existing business, detecting meaningful patterns that indicate unrealized business potential.	Structured and unstructured text and image data from internal and external sources.	• Pattern recognition and analysis • Deep learning • Artificial neural networks	Huge potential profit in a new, untapped field, while competitors must struggle to catch up. (Risks involve reliance on biased or improperly acquired training data.)
Identify untapped market needs and create a better target marketing strategy.	Demographic, psychographic, and consumer purchase behavior data from public and private sources.	• Machine learning • Natural language processing (NLP) • Sentiment analysis • Data mining	Discovery of previously unknown potential business needs and identification of meaningful new niches.
Develop predictive models for complex, global issues, including commerce, public policy, governance, and environmental science.	Structured and unstructured data, probably mostly text, from public and private authorized sources.	• Pattern recognition and analysis • Deep learning • Artificial neural networks	Great public benefit, with likely benefits for companies and other entities acting on these predictions.

WORKBOOK

To identify the primary type of organization you belong to (Optimizer, Extender, Innovator, or Multiplier), it is useful to discuss and then answer some basic questions. These are not meant to be disparaging in any way. It is very possible that your organization is not typical or easily classifiable. However, to better understand how AI projects can be beneficial, please answer as candidly as possible.

Optimizer Questions

On a scale of 1 (not at all descriptive) to 5 (very descriptive), please rate the following characteristics of your organization:

- My organization puts a high priority on reducing costs and/or improving efficiency.

  O 1 O 2 O 3 O 4 O 5

- My organization rarely or never takes extraordinary business risks.

  O 1 O 2 O 3 O 4 O 5

- My organization is a large, well-established player in its market.

  O 1 O 2 O 3 O 4 O 5

- My organization often competes with other companies on the basis of price.

  O 1 O 2 O 3 O 4 O 5

- The prices for my organization's products or services are static or declining.

  O 1 O 2 O 3 O 4 O 5

- My organization often seeks to improve existing products or services but does not often develop new products or services.

  O 1 O 2 O 3 O 4 O 5

Extender Questions

On a scale of 1 (not at all descriptive) to 5 (very descriptive), please rate the following characteristics of your organization:

- In order to grow, my organization frequently seeks to acquire other organizations that provide or develop products and services similar to our own.

  O 1 O 2 O 3 O 4 O 5

- My organization is frequently involved in business mergers with similar organizations.

  O 1 O 2 O 3 O 4 O 5

- My organization's sales force is frequently under pressure to improve its results.

  O 1 O 2 O 3 O 4 O 5

- My organization is always or frequently seeking new ways to increase sales efficiency.

  O 1 O 2 O 3 O 4 O 5

- My organization is always or frequently seeking ways to optimize its CRM or other sales and marketing technology.

  O 1 O 2 O 3 O 4 O 5

Innovator Questions

On a scale of 1 (not at all descriptive) to 5 (very descriptive), please rate the following characteristics of your organization:

- My organization is known for creating new, often unexpected products or services.

  O 1 O 2 O 3 O 4 O 5

- My organization has invested heavily in research and development.

  O 1 O 2 O 3 O 4 O 5

- My organization has strong ties to academic research facilities.

  O 1 O 2 O 3 O 4 O 5

- My organization actively recruits and retains high-level technical talent.

  O 1 O 2 O 3 O 4 O 5

Multiplier Questions

On a scale of 1 (not at all descriptive) to 5 (very descriptive), please rate the following characteristics of your organization:

- My organization is known for defying conventional wisdom.

  O 1 O 2 O 3 O 4 O 5

- My organization is not averse to "bending the rules" in a responsible manner.

  O 1 O 2 O 3 O 4 O 5

- My organization has a history of pursuing entirely new lines of business.

  O 1 O 2 O 3 O 4 O 5

- My organization's leaders actively seek and recruit people who think unconventionally.

  O 1 O 2 O 3 O 4 O 5

After considering these questions, define the implications of being in one of those business categories. What does that mean to an AI project's approach or business objectives?

Sample AI Project Exercise

Once you have determined the basic nature of your organization, it is possible to begin the AI project design process. However, before you do, take the time to analyze a typical process—an ecommerce workflow. There are ten steps in this example, each one requiring a response to user input, an interaction with one or more data systems, or a combination of both. In many cases, the step can potentially be augmented or improved by the application of AI.

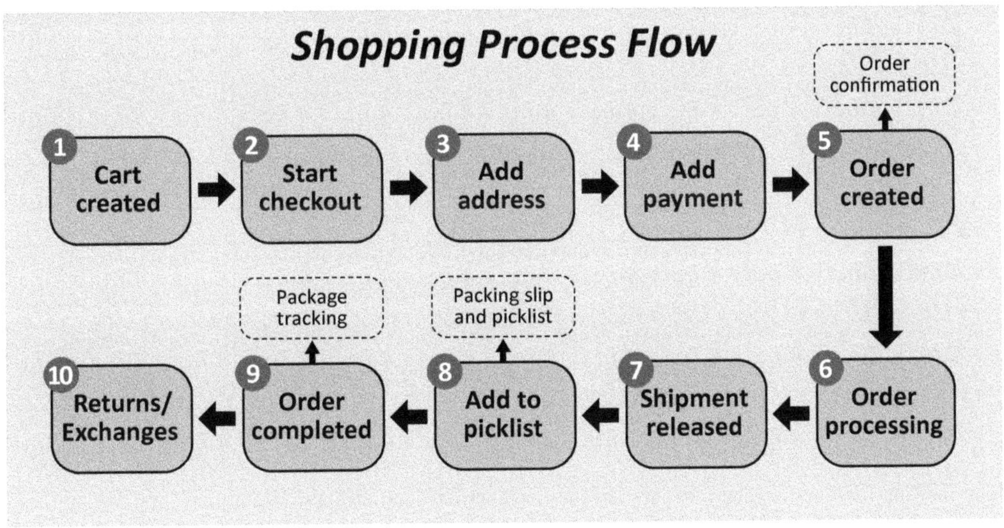

For each step or "node" in this sample workflow, please analyze the diagram and, for each step in the process, write down the particulars as they would be used in an AI project. (Some of the details may become more refined as you become familiar with your organization's data readiness and other factors covered later in the playbook.) These details may include:

- Potential *data sources,* and how the data is to be utilized.
- The *AI processes and goals* that may be applicable at this step,
- The *legal or ethical considerations* involved in using data and AI at this step, and
- What *human interaction or decision* is needed in response to the proposed AI process and how those changes might affect the results. For this example, this can be the customer, the person responsible for handling the physical product, or the person handling or utilizing the resulting data. Can it be automated completely or is a human-in-the-loop needed to ensure that it is appropriate before moving further along in the process?

	Data Source(s)	AI Process/Goals	Legal/Ethical Issues	Human Responses
1				
2				

3				
4				
5				
6				
7				
8				
9				
10				

Your AI Project List

Now that you know what type of organization you belong to, and have practiced analyzing a hypothetical AI project, this is the point where you can list as many potential projects as you or your team can imagine for your type of organization. This has three basic steps:

1. Exercise a ***divergent thinking*** process, otherwise known as a "brain dump." In the chart below, list as many AI projects as you can think of that are appropriate for your type of organization.

General description of the problem to be solved via data & AI	How does It align to a known business goal, directly or indirectly?	What data source(s) are needed and available?	What tangible value does it provide? How will it be measured?

Keep in mind that this list can be as structured or detailed as you wish, but the point is to take an expanded view of *all* the possibilities.

2. Next, after you have completed Section Five of this book, exercise a ***convergent thinking*** process, assigning a priority level to the top three to five projects you have identified here.
3. Finally, after completing the remainder of this book, return to this list and begin to set a timeline, outlining the steps necessary to implement and scale the highest priority AI project in the list.

SECTION TWO

Mastering Data Readiness

In this section, you will learn how to assess and prepare your organization for implementing AI and data strategies. Once you've characterized the basic nature of your organization, the next step is to develop its data readiness, and its relationship with your business goals. Artificial intelligence may not be scary or malevolent, but it's still not easy. As detailed in *The AI Factor,* your data readiness depends on two equally important factors: *organizational maturity* and *internal competence:*

In this section, we will review the steps required for each aspect of a data readiness framework. In the workbook, you will assess your organization's current status, understand our business strategy and align it to your AI Strategy for your business goals, and create a strategy for improving it.

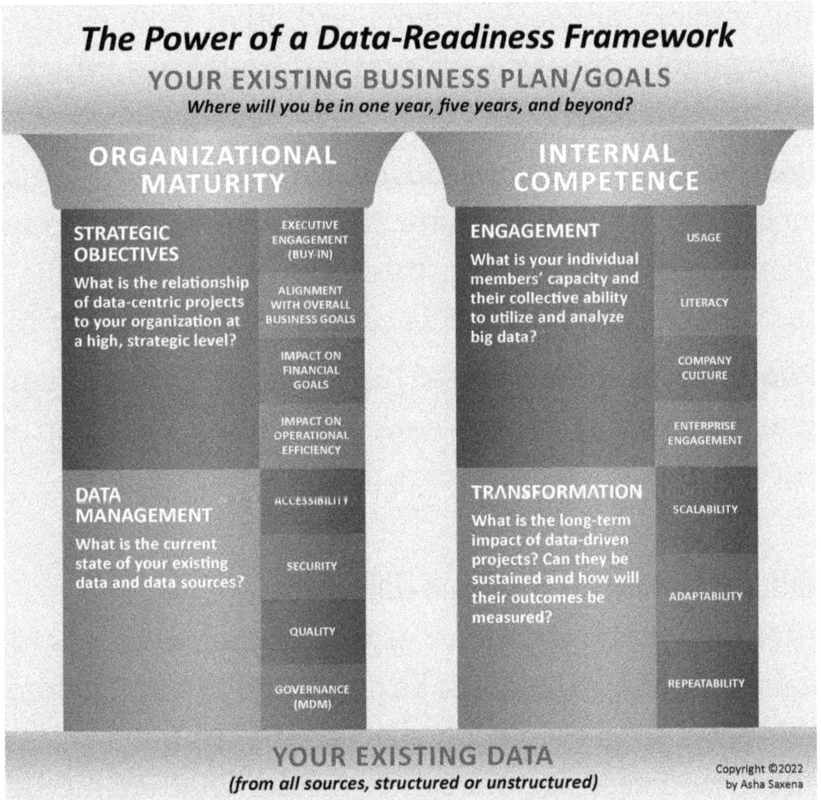

PLAYBOOK

To begin building your AI Strategy, you will need to start with data readiness framework. We need to know if you are ready with a robust data-driven strategy. You must start with the sources of relevant and reliable data. The good news is that most companies have (or at least have access to) much of the data they need, or they learn what additional data is required whether they know it or not. Some potential sources include:

- ***Active data collection***—This involves traditional research techniques, such as surveys, focus groups, and interviews. There are also many third-party data sources from which relevant data may be purchased. It may also involve using AI tools to collect and analyze unstructured data (text, images, and video) from social media and other sources. However, this can raise the issue of "synthetic data" use, which can be problematic.[1]

- ***Passive or permissions-based data collection***—This could include structured purchase data from loyalty programs and ecommerce sites, and unstructured data from public-facing customer reviews, customer service calls, and other sources.
- ***Public data collection***—This includes government data sources, such as census data, containing highly structured but not always current information.

Organizational Maturity—Strategic Objectives

No matter how much data your organization has access to, there are certain key factors that must be taken into account before artificial intelligence projects (or any data-centric initiatives) can hope to be effective. These factors center on the relationship between your projects and your organization as a whole. It involves high-level strategic alignment, not just a casual acknowledgement that AI is "a good idea." In the current public frenzy over AI, it is all too easy to agree without counting the real costs and long-term benefits.

There are three aspects to gauging whether or not your data-centric projects are a good fit with your organization. They are:

- *Executive engagement or buy-in*—In Section Six, we will go into greater detail on the people needed to make AI teams successful. Without substantial buy-in at the C-suite level, organizations will most likely fail to leverage AI in practical ways. Farsighted CEOs will also consider whether or not the company's use of data and AI is ethical and sustainable,

1 Lucini, Fernando. "The Real Deal About Synthetic Data." *MIT Sloan Management Review, 63, no. 1 (Fall 2021): 1-4,* https://www.proquest.com/docview/2600354413.

which will ultimately affect their long-term business success.

- *Alignment with overall business goals*—Of course, this varies according to the optimizer-extender-innovator-multiplier nature of the organization, as covered in Section One. The challenge for traditional IT leadership within an organization is the tendency to focus on historical transactions. However, once the connection is made between the data and the organization's forward-looking goals, it is possible to develop a mindset conducive to successful AI strategy.
- *Impact on financial goals and operational efficiency*—Again, this depends on the type of organization involved. But no matter the organization's growth potential or its willingness to take risks, it must always maintain the relationship between AI strategy and its real-world impact on the business.

Organizational Maturity—Data Management

It goes without saying that all organizations today practice some form of data management, whether the process is explicit or purposeful.[2] Since data is the lifeblood of artificial intelligence, organizations must take a hard look at how well they manage the actual data and their sources.

Data management should always be tied to business strategy, goals, and priorities. A 2021 Gartner Group article pointed out that organizations too often orient data practices around the data rather than business values and outcomes.[3]

---

2 Fryman, Lowell, et al. "The Data and Analytics Playbook." *Science Direct, Elsevier, 2017*, https://doi.org/10.1016/B978-0-12-802307-5.00001-0.

3 Gupta, Ashutosh. "7 Key Foundations for Modern Data and Analytics Governance." Gartner, July 12, 2021, https://www.gartner.com/smarterwithgartner/7-key-foundations-for-modern-data-and-analytics-governance.

The four essential aspects of data management as they pertain to an organization's data readiness are:

- ***Data accessibility***—This begins by answering general questions: "Do we have all the data that we need?" "Do we have access to the relevant data sources?" "Are the data broadly accessible across all levels of the organization?" Great strides have been made in creating practical "dashboards" to visualize important data sources. But if all the data are not freely accessible to those creating AI projects, then they will not have the means to accomplish their business goals.
- ***Data security***—This is a critical factor, as you are working with both internal and external data sources. These may include private data sets of employees, customers, and vendors, as well as government data. "Are your data secure and safeguarded against malicious or improper use?" This question poses the essential dilemma of any AI strategy. Projects must have access to all relevant data, but they must simultaneously protect against malicious actors' use of the original data *and* the results generated by AI. Data privacy, security, and safety includes protecting sensitive information from unauthorized access *and* ensuring compliance with data protection laws (e.g., GDPR, CCPA). It means implementing security measures like encryption, access controls, and regular security assessments.
- ***Data quality***—AI project planners must always determine whether the data is of sufficient quality to create valid business conclusions. The data must be trustworthy, accurate, consistent, and reliable. This involves many factors, including basic accuracy, metadata consistency across multiple data sets, objectivity, and freedom from bias. In practice, this often means that AI project teams must devote significant resources towards data "cleaning," not to alter the data,

per se, but to make sure that AI and machine learning systems have the best possible inputs.

- ***Data governance***—This is an issue framed by a simple question: "Are all your data consistently managed across all your key data sources, or are they siloed and unrelated to each other?" Unfortunately, the answer to this question is seldom simple. In The AI Factor and in many other works, the ideal of *master data management* or MDM is considered to be a worthy goal—and one that is essential to many AI projects. In a nutshell, MDM is based on the idea that there should be a "single version of the truth" for critical business data. Ideally, this involves creating a single master record for each significant entity, item, or event, from multiple data sources and applications—to the point where the information contained in that record is a consistent data definition and a trusted source for making business decisions. It means managing the organization's core business entities (customers, products, employees, etc.) across multiple systems and databases to ensure a single, consistent view of these key data elements. It also means managing metadata consistently to improve understanding and usage of data assets. This includes documenting the data's definitions, sources, transformations, and lineage to enhance transparency and trust in data.

Clearly, all four elements are part of any organization's ongoing IT challenge, made even more challenging with the growth of unstructured data. But as strategic AI projects gain momentum and credibility, AI itself may become a means of improving an organization's data management practices.

Internal Competence—Engagement

The second "pillar" of an organization's data readiness framework involves its internal competence, both in terms of its people as well as its capacity to sustain initiatives. The first part, engagement, is covered in greater detail in Section Four, but the basic principles are as follows:

- ***Data use***—With any data-centric initiative, the key question must be, "Are key individuals able (and willing) to view, comprehend, and utilize relevant data from across the organization?" Most of us need help making sense of qualitative data, especially when there is so much to consider. There are many tools for data visualization, of course, including those created by AI. However, in order to develop new AI and data initiatives, team members must have the tools and the skill sets to make sense of their raw materials.
- ***Data literacy***—As we will explore in Section Four, not all members of an AI project team are data specialists, nor should they be. However, all such non-IT individuals should understand the basic nature and value of data. They should also be able to distinguish between the hyperbole surrounding AI and big data from the practical reality they represent to the organization. As the conclusion of an MIT Sloan report stated, "True data literacy should enable one to think and act differently—start by understanding the real business problem and use intelligent insights to solve the right problems."[4]
- ***Company culture***—This is an over-used term, but in this context it refers to an organization's overall receptivity to using data—not only to describe what happened in the

4 Brown, Sara. "How to Build Data Literacy in Your Company." *MIT Sloan Management Review, February 9, 2021,* https://mitsloan.mit.edu/ideas-made-to-matter/how-to-build-data-literacy-your-company.

past but also to plan future actions and reinvent business models. Although a robust data culture often starts with executive-level buy-in, it must also have grassroots support throughout the company. That means data-centric projects cannot be framed as "cool science experiments," or just collecting data for data's sake. Rather, they must be deployed to make good decisions—and the results of those decisions made known throughout the company. AI also presents challenges in the realm of change management.[5] Using AI just for its own sake, simply as a replacement for human input, may indeed create some short-term benefits, but, if done carelessly, may create costly long-term problems. You should always consider what types of change management will be needed to adapt the organization and the staff to the new ways of working, or gain comfort with the new skills to use the new technology.

- ***Enterprise engagement***—This is the most aspirational, but still essential aspect of this data-readiness framework. Based on a philosophy of "stakeholder capitalism," it asks the question, "Is the organization committed to including all stakeholders, internal and external, as integral to its data strategy?" Data analytics and AI are critical components of enterprise engagement. To make decisions that provide value to *all* stakeholders—including a company's employees, partners, and customers—it must first know them, and have a clear understanding of their habits, their preferences, and the likelihood of their future actions. That knowledge can only be acquired and applied by understanding and leveraging big data.

---

5 Saha, Debanjan. "Navigating Change Management in the Era of Generative AI." *Forbes*, *5 Oct. 2023*, www.forbes.com/sites/forbestechcouncil/2023/08/17/navigating-change-management-in-the-era-of-generative-ai/.

Internal Competence—Transformation

There is nothing worse for a data-centric initiative than the idea that it was a one-time success. It takes effort to gather the right data, ask the right questions, and use machine learning and/or deep learning techniques to seek relevant patterns in the data, in order to execute faster processes or actions. If all that effort has to be done all over again the next time, it will discourage, or even halt, future attempts. This is true even if the results are good for the organization. Big data and AI can only be transformational if the company can build on past successes with relative ease.

To achieve this, organizations must possess and utilize the tools necessary to turn each AI project into part of a greater whole. These include the ***scalability of one's initial project, its adaptability to other business problems, and its repeatability over time. While each of these factors speaks to your organization's data readiness (and especially to its internal competence), the details of these three factors will be discussed in Section Seven.***

WORKBOOK

A sound data readiness framework is foundational to all artificial intelligence and big data projects. In this section, consider how well your organization meets these criteria, how it may impact your sample project, and what steps could be taken to remedy or improve your overall data readiness.

Strategic Objectives Questions

On a scale of 1 (unknown or not at all descriptive) to 5 (very descriptive), please rate the following characteristics of your organization, especially as it relates to the AI project you outlined in Section One:

- Our top management consistently use data to meet or adjust targets, manage change, and make data-driven business decisions.

  O 1 O 2 O 3 O 4 O 5

- Our top management is committed to AI strategy, including a clear strategic plan, clear AI objectives and budget, and clear communication with shareholders.

  O 1 O 2 O 3 O 4 O 5

- Our top management has a good track record of making business decisions based on reliable data and the intelligent application of AI.

  O 1 O 2 O 3 O 4 O 5

- Our top management understand the true potential of AI and are aware of its requirements, limitations, and the importance of sustainable best practices.

  O 1 O 2 O 3 O 4 O 5

- Our top management has clearly communicated their vision of AI to employees, suppliers, and customers.

  O 1 O 2 O 3 O 4 O 5

- The focus of our AI strategic plans are clear in the daily activities of our management.

  O 1 O 2 O 3 O 4 O 5

- In general and for AI and data projects, performance expectations are set on an individual, team, and organizational level.

  O 1 O 2 O 3 O 4 O 5

- In general and for AI and data projects, there is a performance management system in place that measures and identifies competence, skills, and training needs.

  O 1 O 2 O 3 O 4 O 5

- In general and for AI and data projects, the rewards and consequences for performance are clearly outlined, and there are regular communication forums in place to ensure transparency.

  O 1 O 2 O 3 O 4 O 5

- This AI project defined in Section One is well aligned with our organization's overall, long-term business goals.

  O 1 O 2 O 3 O 4 O 5

- This AI project defined in Section One will have a measurable, positive impact on our financial goals.

  O 1 O 2 O 3 O 4 O 5

- This AI project defined in Section One will have a measurable, positive impact on our operational efficiency.

  O 1 O 2 O 3 O 4 O 5

If you wish, you can average the scores, but the nature of your strategic objectives should become obvious from your answers. To qualify these, please take time to describe the details. Also, on your own or from a group discussion, note the possible ways in which these factors can be improved:

Strategic Objectives SUMMARY			
Executive Engagement	**Business Alignment**	**Financial Impact**	**Efficiency Impact**
Who are the C-level execs involved and how committed are they?	To what specific business strategy is the AI project aligned?	What is the tangible, measurable financial outcome?	What is the tangible, measurable efficiency outcome?
NOTES			

Data Management Questions

On a scale of 1 (unknown or not at all descriptive) to 5 (very descriptive), please rate the following characteristics of your organization, especially as it relates to the AI project you outlined in Section One:

- In general, all of our business data are broadly accessible across all levels of our organization.

  O 1 O 2 O 3 O 4 O 5

- In general, all of our business data are secure and protected against malicious or improper use.

  O 1 O 2 O 3 O 4 O 5

- In general, we are aware of the specific data that would be compromised in the event of a data breach, and we are fully prepared to respond in such an event.

  O 1 O 2 O 3 O 4 O 5

- In general, we have a clear distinction between company-owned data and user data, and we have the proper level of consent in place to collect customer data.

  O 1 O 2 O 3 O 4 O 5

- In general, our organization's business data are not siloed and unrelated, but are well managed across all domains, using MDM or its equivalent.

  O 1 O 2 O 3 O 4 O 5

- In general, we are well aware of how much of our data is stored in the cloud, and who has access to our various data assets.

  O 1 O 2 O 3 O 4 O 5

- In general, our data supports the fulfillment of our organization's purpose.

  O 1 O 2 O 3 O 4 O 5

- In general, our data supports the achievement of our organization's business strategy.

  O 1 O 2 O 3 O 4 O 5

- For this particular AI project, the required data—*and* the resulting data—are both fully accessible across the organization and protected against malicious or improper use.

  O 1 O 2 O 3 O 4 O 5

- For this particular AI project, the required data are of sufficient quality and consistency to create valid business conclusions.

  O 1 O 2 O 3 O 4 O 5

- For this particular AI project, the required training data—and the resulting business data—are well managed across all relevant domains.

  O 1 O 2 O 3 O 4 O 5

If you wish, you can average the scores, but the nature of your data management practices should become obvious from your answers. To qualify these, please take time to describe the details. Also, on your own or from a group discussion, note the possible ways in which these factors can be improved:

Data Management SUMMARY			
Data Accessibility	**Data Security**	**Data Quality**	**Data Governance**
Which data sources do we need to access?	What data security, policy and regulation needs to be in check?	What data quality principles need to be followed?	How will good data governance policies be initiated, confirmed, and followed?
NOTES			

Engagement Questions

On a scale of 1 (unknown or not at all descriptive) to 5 (very descriptive), please rate the following characteristics of your organization, especially as it relates to the AI project you outlined in Section One:

- Across our organization, key individuals are able and willing to view, comprehend, and utilize relevant data from all available sources.

  o 1 o 2 o 3 o 4 o 5

- Across our organization, most or all non-IT members have a reasonably high level of understanding of the basic nature and value of data.

  o 1 o 2 o 3 o 4 o 5

- Across our organization, all members are consciously aware of their stake in the use of data and AI to advance their shared interests.

  o 1 o 2 o 3 o 4 o 5

- For this particular AI project, all team members—including those outside IT— are willing and able to utilize relevant data from any available source.

  o 1 o 2 o 3 o 4 o 5

- For this particular AI project, all team members—including those outside IT—have a high level of *general* data literacy.

  o 1 o 2 o 3 o 4 o 5

- For this particular AI project, all team members are open to unexpected outcomes that run counter to their previous opinions.

  o 1 o 2 o 3 o 4 o 5

If you wish, you can average the scores, but the nature of your engagement level should become obvious from your answers. To qualify these, please take time to describe the details. Also, on your own or from a group discussion, note the possible ways in which these factors can be improved:

Engagement SUMMARY			
Data Usage	**Data Literacy**	**Company Culture**	**Enterprise Engagement**
Are those involved in the project able to access all the relevant data?	Who among those involved in the project are sufficiently data literate?	What is the company's overall acceptance of using data and AI to plan future actions?	Do *all* the company's stakeholders recognize the value of leveraging big data and AI?
NOTES			

SECTION THREE

Sustainable, Responsible, and Ethical AI

Other AI playbooks usually include a section on making sure artificial intelligence initiatives are ethical, responsible, and explainable, avoiding wherever possible the negative long-term consequences of data and AI use. However, this material can be located towards the end of the "how-to" material, creating the appearance of an afterthought. So, in order to implement an AI strategy in a truly holistic, responsible manner, one must do so with a *prior* understanding of the frameworks for ensuring its sustainability.

PLAYBOOK

Artificial intelligence and the use of big data can multiply your business success. But if used irresponsibly, illegally, or unethically, they can also multiply the risk and your chances of *failure*. AI has tremendous power—both positive and negative—that will affect the people with whom you do business and the planet where your business needs to thrive.

AI sustainability has many facets, each requiring buy-in by the entire team as well as the means of ensuring responsible practices. The broad categories of AI sustainability can be summarized as follows:

- ***AI and big data projects must be explainable and transparent,*** especially those with autonomy or the capacity to self-improve. This has three parts:
    - ***There must be satisfactory explanations of an AI system's actions,*** auditable by competent human authority.
    - ***If an AI causes harm, the cause must be discoverable*** and the means of correcting that harm be made clear.
    - ***Sustainable AI is not a "black box."*** Its workings and logic should always be understood and manageable by qualified human beings.
- ***AI and big data projects must be ethical***—This means several things, which will be covered later in the playbook. They include:
    - ***Justice and fairness***—in accord with shared values of human dignity, autonomy, and diversity
    - ***Non-maleficence***—not designed with a destructive purpose, but rather with safety, security, and of benefit to as many people as possible
    - ***Responsibility***—for all those who design, build, and use AI systems to have a practical stake in their use, misuse, and results

  - ***Privacy and choice***—always maintaining the rights of the humans who generated the data used by an AI system

- ***AI and big data projects must be legal***—This means proactive adherence to the spirit of AI legal requirements established to protect individual rights and prevent harm. This should be done not only to avoid legal liability, but also to establish an earned reputation that will promote your long-term success.

Explainability Explained

All artificial intelligence and big data projects affect all human beings and the organizations to which they belong. However, the reality is that most people do not understand what AI actually is. We are susceptible to narrative myths created by fictional works like *2001: A Space Odyssey* or *The Terminator*. As a species, we have evolved to react subconsciously to the unknown or the unfamiliar.

Artificial intelligence today is definitely in that category, making it all the more important to present AI as a normal aspect of our lives—a tool like any other that can be used for good or ill. This was the basis for the first part of *The AI Factor*, which identified a number of areas you will want to consider, and explore their implications for your company, its AI initiatives, and the extent to which you may need to address or mitigate them:

- ***AI is ubiquitous.*** From voice assistance to maps to chatbots to streaming video, AI has become a common and often welcome part of our lives.
- ***AI is invisible.*** We are not always aware of what AI is doing in the background—any more than we are aware of other technologies that can, at least potentially, make our lives better.

- ***AI is hard to explain.*** But so are other technologies we have become used to, like the workings of your car. However, like a car, AI and machine learning **can** be understood in general, non-technical terms—to the point where it is no longer a perceived obstacle.
- ***AI is fallible.*** Like all tools, especially sophisticated ones, AI can go wrong in unusual or unpredictable ways. Owning that reality, and being prepared to address it in our projects, is the foundation of all sustainable work in AI.

Besides explaining AI in general, it is incumbent on AI project teams to make their aims and methods as clear as possible to non-technical stakeholders. As we will explore in Section Three (data readiness), this includes:

- ***Decision makers*** at the top of an organization, who must understand the benefits and potential risks before they buy in to an AI project.
- ***Domain knowledge experts*** whose *expertise* outside the mechanics of AI is essential to a project's success.
- ***Those responsible for monitoring results*** and responding to them.
- ***Those whose lives will be impacted*** by the project. The clearer an AI project's aims and methods are, without revealing protected intellectual property, the more sustainable and beneficial it will be in the long run.

In 2020, data science and AI researcher Dr. Joyjit Chatterjee pointed out that, "good AI isn't just accurate, it's transparent and scalable," and thereby trustworthy by definition.[1] He presented a basic framework for achieving transparency, including:

---

1 Chatterjee, Dr. Joyjit. "AI beyond Accuracy: Transparency and Scalability." *Medium, Towards Data Science, 14 May 2020,* https://towardsdatascience.com/ai-beyond-accuracy-transparency-and-scalability-d44b9f70f7d8.

- Using simple and easy-to-use libraries for explainable AI.
- Identifying and using *causal* relationships within your AI model, rather than rely solely on *correlation*—which is easier to compute but may not tell the whole story.

Chatterjee encouraged AI teams to simplify projects, reducing redundant features, and using simpler data models wherever possible. This, he asserted, will result in not only AI models with greater explainability but also more potential to scale.

In the workbook part of this section, we will help you practice ways in which an AI project can become less of a black box approach and more of a practical, explainable business tool.

Models for AI Explainability

Many AI models, including semi-autonomous machine learning approaches and those involving deep learning-based systems, cannot be easily explained to non-scientists. This results in these systems being more opaque than necessary, and more likely to create liability. However, there are three ways that organizations can explain their AI without mathematics and without revealing an organizations' valuable intellectual property:

- ***The Inferred Explanation Approach*—**This is the easiest method to deploy and the one best suited to provide a top-level understanding to decision makers. It involves asking questions designed to produce *correlations* between the data set and what the AI is designed to do. These questions can be case-based ("Is there a previous situation that led to comparable results?") or merely based on accepted expert knowledge ("Why does this result make sense?"), among others.[2] The problem, of course, is that *correlation is not*

2 Chari, Shruthi, et al. "Explanation ontology: A general-purpose, semantic representation for supporting user-centered explanations." *Semantic Web, 2023, pp. 1–31,* https://doi.org/10.3233/sw-233282.

causation. While it provides a reasonable, human-understandable explanation, this method does not get into the AI black box itself.

- ***The Feature Extrapolation Approach***—This is a more time-intensive method, requiring the AI team to state whether or not (and to what extent) the system relies on extrapolating conclusions from unstructured data in the training or machine learning process. This is increasingly used when the training data consist of images, such as those used in healthcare. Extrapolation from unstructured data is not a bad thing in itself, and is often essential to machine learning. The need is for greater transparency when extrapolation is used, so that domain experts can evaluate whether or not it generates valid results.[3]
- ***The Key Variable Analysis Approach***—This provides the most precise explanation of an AI system, but it requires a full statistical analysis of each input on the system's decision-making process. It is the most effective way to detect problems of bias in the data. However, besides being time-consuming, it can produce findings that are less clear to non-scientists. It may also reveal aspects of your system to unauthorized or malicious actors.

Steps for Assuring AI Equity

In 2022, the World Economic Forum issued a "Blueprint for Equity and Inclusion" as it relates to AI projects.[4] The white paper outlines a valuable, seven-step model for creating sustainable artificial intelligence projects. At each step, it combines the

3 Cao, Xuenan, and Roozbeh Yousefzadeh. "Extrapolation and AI transparency: Why machine learning models should reveal when they make decisions beyond their training." *Big Data & Society*, vol. 10, no. 1, 2023, https://doi.org/10.1177/20539517231169731.

4 "A Blueprint for Equity and Inclusion in Artificial Intelligence." World Economic Forum, 29 June 2022, www.weforum.org/publications/a-blueprint-for-equity-and-inclusion-in-artificial-intelligence/.

responsibilities of AI developers or builders with those of a project's managers and stakeholders.

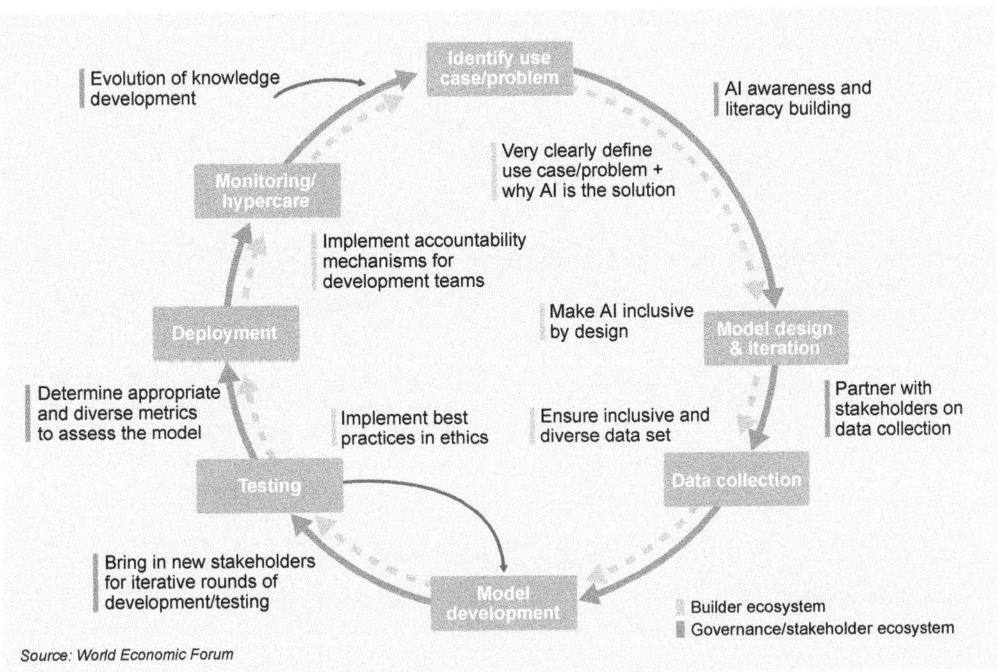

Source: *World Economic Forum*

In Sections Two and Three, we outlined the criteria for prioritizing an AI use case or problem. It must represent a significant business value and/or resolve a significant business problem. Also, the data for such a project must be of sufficient quality, and the organization itself must be sufficiently *data ready*. But its impact on equity and inclusion must also be considered in order for it to be viable in the long term.

At the beginning phase of the project, its managers should be focused on raising general AI awareness and literacy. Besides basic education and awareness-building (which should be company-wide where possible), project managers should always attempt to define the problem from the perspectives of more than one demographic group, and determine if the problem impacts one group differently than another.

Another practice that executives and managers need to apply at this stage—and throughout the process—is to maintain hiring practices conducive to equitable representation within AI teams.

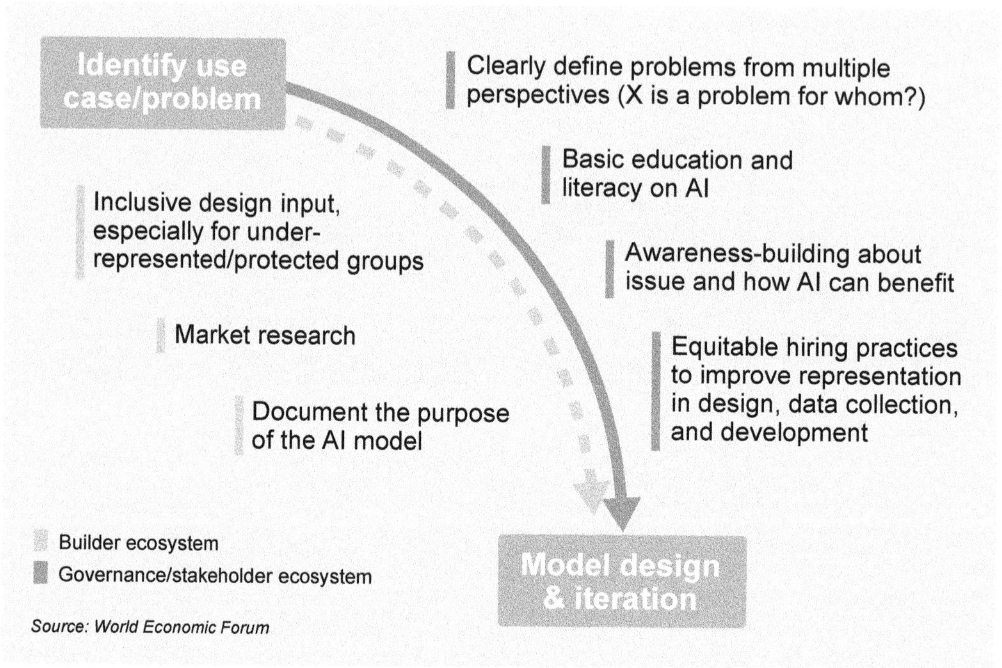

On the builder side, clearly defining the problem, and how AI can solve it, leads to the actual design of the AI model. Here the team should always include design input from a broad range of groups, especially those who are traditionally under-represented. Documenting the purpose of the AI model is essential, as it contributes to overall transparency.

Between the initial model design and final model development, the project team must focus on data collection and evaluation. On the developer side, identifying data needs is an obvious requirement, as is the need to include data from demographically diverse sources. Just as the builder side needs to partner with stakeholders in the design process, so too must the project's managers partner with stakeholders on maintaining diverse data collection methods and sources.

Once the AI model is created, testing involves equally important practices to ensure equity and inclusion. On the governance side, managers should always include new stakeholders in the reiterative development and testing process—to reduce the chances of unintentional bias. During the testing and development "loop," the data engineers and testers must also follow a reasonable set of best practices with regard to ethics, transparency, and privacy—as described earlier in this section.

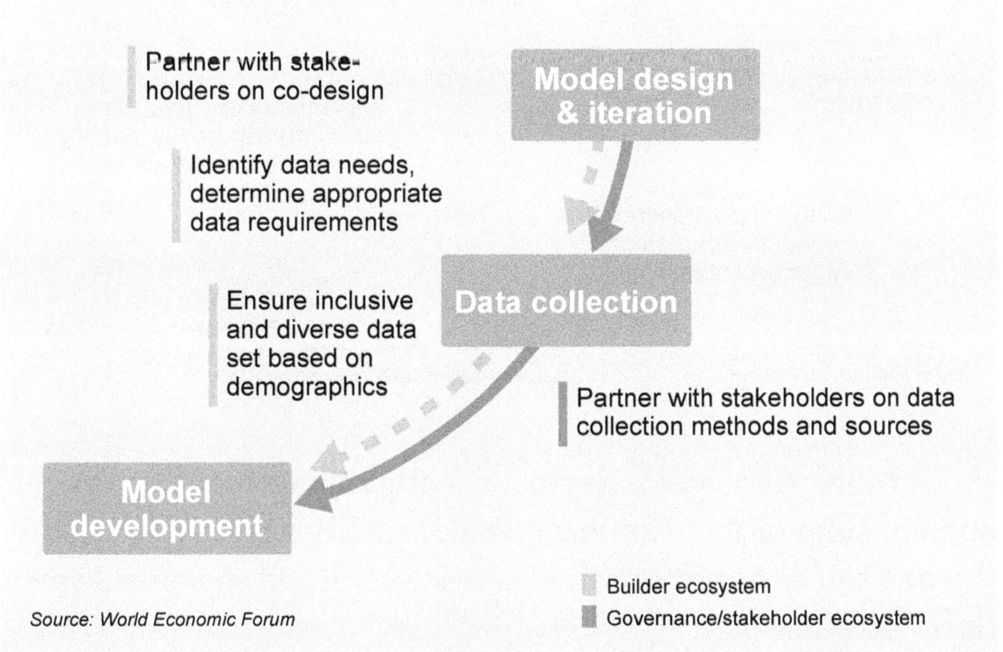

Source: World Economic Forum

Following the testing phase, of course, the AI project must be deployed. From a diversity and inclusion perspective, the governance side of the team should always measure the impact of the project's results on a wide array of communities and demographic groups. Over time, it should also develop appropriate metrics for assessing AI project impacts on such groups.

The technical developers have many responsibilities at this stage, starting with technical verification, user acceptance testing, and debugging. Beyond that, they should also develop a beta testing approach that includes a broad array of demographic groupings. As part of the test feedback process, they should also implement accountability mechanisms to gather feedback on possible diversity and inclusion-related issues with the project.

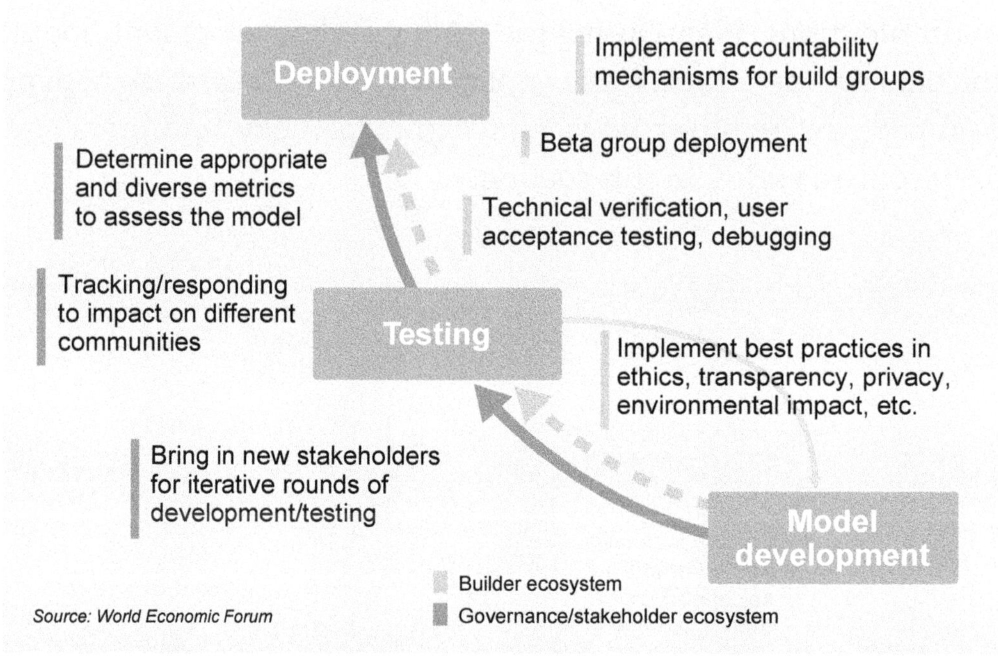

Source: World Economic Forum

Once the AI model is deployed, both management and development sides of the team have responsibilities related to making this and future AI projects diverse, inclusive, and sustainable over the long term

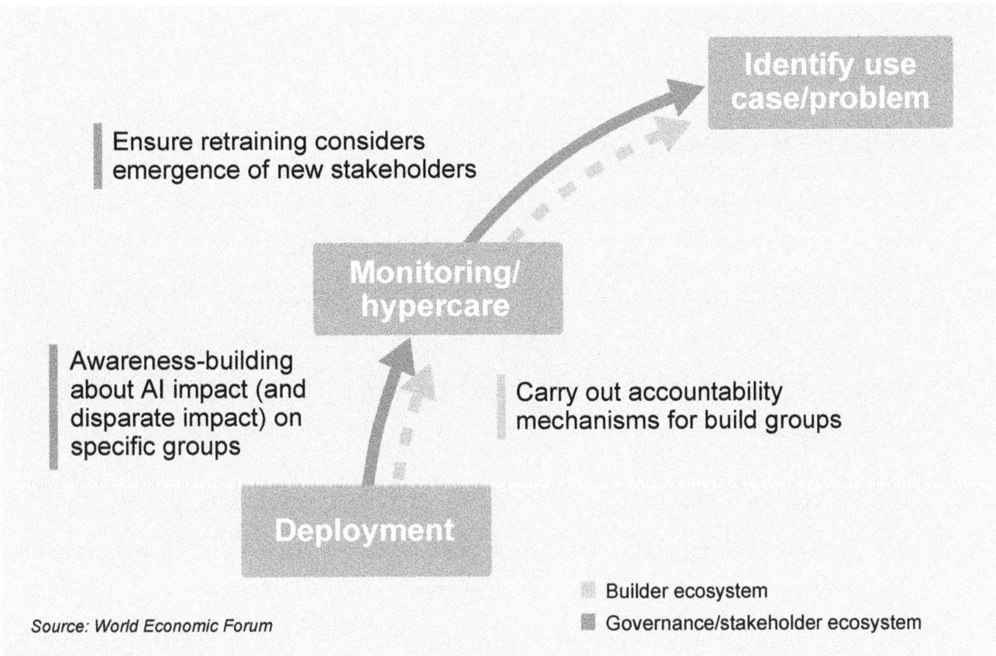

Source: World Economic Forum

Developers must consistently carry out the accountability mechanisms they created during the testing and deployment stages. Managers need to focus on ongoing awareness-building on AI's impact, both good and bad, on specific demographic groups. As the project cycles towards further AI development, they should continuously seek out new stakeholders that the project may have missed or undervalued.

Doing the Right Thing—Even If No One Is Looking

The late NYU professor Neil Postman once said, "Every technology is both a burden and a blessing; not either-or, but this-and-that." This is true of artificial intelligence, but the difference is that the blessings ***and*** the burdens of AI happen much faster than they do with previous technology.

The rapid growth of AI and machine learning has resulted in some spectacular failures, prompting many organizations to develop comprehensive ethical guidelines. Governments too have accelerated their regulatory efforts, as we'll see later in this playbook. ***But ethical frameworks alone are not enough.*** Each organization and team must start with a commitment to basic rules, including:

- ***Use data wisely.*** The past two decades have seen an unprecedented use of private user data for the benefit of tech companies and their advertising partners. Public backlash against this has been overwhelming, and is only likely to grow. GDPR compliance, the potential for lawsuits, and sluggish but inevitable government regulation are only the beginning.

  Big data availability may also no longer be a given. With advances in secure blockchain and other Web3 technologies, formerly "free" user data will be increasingly hard to come by, as more people begin to assert their privacy rights and use new technology to protect them. So, in order to conserve the long-term viability of AI initiatives, organizations must find ways to acquire and use data, even by paying for it, from viable, privacy-honoring sources.
- ***Always be aware of the potential for data bias.*** Data about people and their activity come from human sources. So, one must always ask probing questions about who those humans are, and whether or not they are biased in any way. A group of mostly-white, mostly-male employees will likely create a training data set that skews towards similar individuals and excludes others. The same problem holds true for other, less structured data. Data from images of cancerous versus non-cancerous tissue samples can be skewed, for example, if one group of images included a measurement scale and the other did not. *AI is only as good as the data it sees.*
- ***Avoid situations that offer a short-term benefit to a small group while ignoring or harming other groups.*** It can be tempting to take shortcuts that offer quick rewards for those who act first. With many AI initiatives, as with any other

"get-rich-quick" proposition, such promises are short-lived at best. In the long run, the best AI projects are those with a constructive purpose involving the safety, security, and other benefits (including financial) of as many people as possible.

- Own your AI initiatives. Making your team (and ultimately yourself) accountable for an AI project's successes—and it's failures—is the greatest incentive for AI sustainability. This is no different from any other healthy business practice, except that with AI, the rewards and the accountability for mistakes will happen much more rapidly.

The cost of ***not*** implementing sustainable AI practices far outweighs the cost of doing so. As AI and analytics law practitioner Andrew Burt noted in 2020,[5]

> **"[T]he task of translating AI principles into concrete metrics is both intellectually challenging and resource intensive. ...[But] the alternative—waiting to measure the harms of AI until after they occur—will be far more difficult and costly to their consumers, their reputations, and their bottom lines.**
>
> **"Ultimately the question is not *if* AI harms need to be measured but *when*. If organizations wait too long to quantify the damages their AI can cause, the courts will start doing it for them."**

Government Oversight of AI

The rapid growth of AI development has far outpaced the efforts of governments to regulate it and curtail its abuses. Early efforts at regulation have come primarily from the European Union, the United Kingdom, Canada—and belatedly from the United States. However, US companies should not become complacent. Any company that does significant business overseas should be aware

5 Burt, Andrew. "Ethical Frameworks for AI Aren't Enough." *Harvard Business Review, 9 Nov. 2020,* https://hbr.org/2020/11/ethical-frameworks-for-ai-arent-enough.

of governments' increased scrutiny of AI practices, especially as they relate to privacy issues.

Using the EU as a model, government action affecting AI initiatives will likely fall into three categories:

- ***Prohibition of AI that carries unacceptable risk***—This may include cognitive behavioral manipulation, classifying people based on behavior, socio-economic status, or personal characteristics. It may also include biometric identification systems, such as facial recognition. At this level, strict enforcement will likely be limited to companies doing business in EU countries, with enforcement becoming less likely in isolated or authoritarian countries.
- ***Strict regulation of high-risk AI systems***—This may include some of the above areas, and, in addition, systems affecting public safety, critical infrastructure, education and training, HR and worker management, law enforcement, access to public services, migration and border control, among others areas of public concern.
- ***Requirements for generative AI and limited risk systems***—While likely not becoming prohibited, these systems will be required to disclose a work's AI-generated status. Compliant systems will be designed to prevent the creation of illegal or harmful content, and to disclose summaries of copyrighted data used for training. Limited risk AI systems will at least require transparency, giving AI users and content consumers the ability to make informed decisions.

A summary of the current state of AI regulation, led by initiatives in the EU, can be summarized as follows:

- ***The European Union***—The EU's Artificial Intelligence Act[6] was passed in December 2023.[7] Although there will be a two-year grace period, allowing companies time to comply, the fines for companies using AI in "high risk" situations (affecting safety or fundamental rights) could face penalties and fines as high as $10 billion.

  The European Parliament's priority is "to make sure that AI systems are safe, transparent, traceable, non-discriminatory, and environmentally friendly. AI systems should be overseen by people, rather than by automation, to prevent harmful outcomes."[8]

  As with the EU's earlier framework for data protection and privacy, the General Data Protection Regulation, or GDPR, companies doing business in Europe will have to include consideration of their AI framework in order to remain competitive.

- ***The United Kingdom***—Across the Channel, AI regulation has taken a slightly different, but still serious form. The country's four digital regulators have combined under the banner of the Digital Regulation Cooperation Forum, with the Information Commissioner's Office, or ICO, taking a lead role when it comes to AI. The ICO provides regulatory and practical guidance to companies doing business in the UK.[9]

---

6 European Commission, Directorate-General for Communications Networks, Content and Technology. "Proposal for a Regulation of the European Parliament and of the Council Laying Down Harmonized Rules on Artificial Intelligence (Artificial Intelligence Act) and Amending Certain Union Legislative Acts." CNECT, 21 April 2021, https://eur-lex.europa.eu/legal-content/EN/TXT/?uri=CELEX:52021PC0206.

7 Satariano, Adam. "E.U. Agrees on Landmark Artificial Intelligence Rules." *The New York Times, 8 Dec. 2023,* https://www.nytimes.com/2023/12/08/technology/eu-ai-act-regulation.html.

8 "EU AI Act: First Regulation on Artificial Intelligence." *News: European Parliament,* 14 June 2023, www.europarl.europa.eu/news/en/headlines/society/20230601STO93804/eu-ai-act-first-regulation-on-artificial-intelligence.

9 "Artificial Intelligence." ICO, https://ico.org.uk/for-organisations/uk-gdpr-guidance-and-resources/artificial-intelligence/. Accessed 26 Oct. 2023.

- ***Canada***—In 2022, Canada enacted the Artificial Intelligence and Data Act (AIDA),[10] imposing a risk-based framework similar to that of the EU, as well as criminal law prohibitions against reckless or malicious use of AI.
- ***The United States***—There is currently no comprehensive US framework or legislation specifically involving artificial intelligence. However, various federal agencies have taken action to curb abuses related to AI's impact on privacy, employment discrimination, and other areas of public concern. One of the more daunting measures in this regard is the existing Federal Trade Commission (FTC) model *deletion tool*,[11] which would force companies using AI illicitly to delete their algorithms, along with all data collected with those algorithms. Since 2019, the agency has used this tool four times, including two actions against Amazon.

  US reluctance to advance meaningful AI legislation has many explanations, most often attributed to political influence by large tech companies. There is growing bipartisan support for responsible AI use, however. In 2020, the previous administration issued an Executive Order promoting the use of trustworthy AI within the US government.[12]

  More recently, the current administration issued a proposed "AI Bill of Rights,"[13] followed by an Executive Order, emphasizing the need to protect individual rights from the misuse

10 "The Artificial Intelligence and Data Act (AIDA)—Companion Document." Government of Canada / Gouvernement du Canada, 13 Mar. 2023, https://ised-isde.canada.ca/site/innovation-better-canada/en/artificial-intelligence-and-data-act-aida-companion-document.

11 Riley, Tonya. "The FTC's Biggest AI Enforcement Tool? Forcing Companies to Delete Their Algorithms." CyberScoop, 8 July 2023, https://cyberscoop.com/ftc-algorithm-disgorgement-ai-regulation/.

12 "Promoting the Use of Trustworthy Artificial Intelligence in the Federal Government." *Federal Register*, 3 Dec. 2020, www.federalregister.gov/documents/2020/12/08/2020-27065/promoting-the-use-of-trustworthy-artificial-intelligence-in-the-federal-government.

13 "Blueprint for an AI Bill of Rights." The White House, The United States Government, 16 Mar. 2023, www.whitehouse.gov/ostp/ai-bill-of-rights/.

> of AI.[14] Similar proposals in Congress, while slow-moving, will eventually add "teeth" to these principles, and align the United States with the EU when it comes to enforcing responsible AI practices.

The move to more sustainable AI is not limited to Europe and North America. Worldwide initiatives and conferences, such as The AI Summit, are receiving support and attention from China and other countries—as well as developers at the forefront of AI. Whether your project is large or small, transformative or simply a move towards greater efficiency, making your AI sustainable is not an option.

WORKBOOK

Nearly all organizations developing or adopting AI are on record supporting its ethical and responsible use. So, this workbook will devote less time on the public policy or public relations aspects of AI sustainability and more on the practical steps needed to ensure that AI projects *are*, in fact, sustainable. Wherever possible, these exercises are aimed at achieving best practices now—*before* regulations or threats of legal liability compel organizations to do the right thing.

In the next section, we will explore your organization's data readiness—the prerequisite for undertaking artificial intelligence projects as a routine business practice. However, before doing so, your organization's practical understanding of explainable, ethical, and legally compliant AI must be well established. Following that, the workbook will address the sustainability issues involved in the project you developed in Section One.

---

14 "Fact Sheet: President Biden Issues Executive Order on Safe, Secure, and Trustworthy Artificial Intelligence." The White House, The United States Government, 30 Oct. 2023, www.whitehouse.gov/briefing-room/statements-releases/2023/10/30/fact-sheet-president-biden-issues-executive-order-on-safe-secure-and-trustworthy-artificial-intelligence/.

Organizational Awareness & Accountability Questions

On a scale of 1 (unknown or not at all descriptive) to 5 (very descriptive), please rate the following characteristics of your organization:

- My organization already has existing infrastructure (on data governance, compliance, privacy, and other data-related risks) that can be applied to AI strategy and practices.

  O 1 O 2 O 3 O 4 O 5

- My organization has explicit and enforceable policies regarding the proper use of personal data for AI projects.

  O 1 O 2 O 3 O 4 O 5

- My organization has a data and AI ethical risk framework appropriate to my industry (e.g., emphasizing privacy issues in healthcare, associative bias in retail, racial or gender bias in recruiting).

  O 1 O 2 O 3 O 4 O 5

- My organization has a standard procedure or methodology for making its AI initiatives explainable to decision makers and auditable by qualified domain experts.

  O 1 O 2 O 3 O 4 O 5

- The executives and managers responsible for AI initiatives have clear and public accountability for the success or failure of the company's AI projects.

  O 1 O 2 O 3 O 4 O 5

- Within their respective departments, members of AI-related teams have clear lines of reporting and accountability for their roles in AI projects.

  O 1 O 2 O 3 O 4 O 5

AVERAGE SCORE ____________

While company AI policies and guidelines alone do not guarantee that your project will be ethically and legally sustainable, a high average score will set a reasonable baseline. A low score means your organization has a lot of work to do.

AI Explainability Questions

On a scale of 1 (unknown or not at all descriptive) to 5 (very descriptive), please rate the following characteristics of the AI project you outlined in Section Three:

- My AI project includes a well-defined procedure for describing the data sources used by our AI initiatives to non-technical subject matter experts and decision-makers.

  ○ 1 ○ 2 ○ 3 ○ 4 ○ 5

- My AI project includes a well-defined procedure for explaining its actions in a manner that can be audited by a qualified, non-technical expert.

  ○ 1 ○ 2 ○ 3 ○ 4 ○ 5

- My AI project includes a well-defined procedure for discovering the reasons why it may cause harm, and for making clear all the possible remedies.

  ○ 1 ○ 2 ○ 3 ○ 4 ○ 5

- My AI project's explanations are more about causal effects than about correlations.

  ○ 1 ○ 2 ○ 3 ○ 4 ○ 5

- My AI project includes human-readable feedback that would give qualified domain experts the ability to monitor its impact and effects.

  ○ 1 ○ 2 ○ 3 ○ 4 ○ 5

If you wish, you can average the scores, but your project's potential explainability should become obvious from your answers.

AI Data & Ethics Questions

On a scale of 1 (unknown or not at all descriptive) to 5 (very descriptive), please rate the following characteristics of the AI project you outlined in Section Three:

- The AI project's goals and methods are in accord with shared values of human dignity, autonomy, and diversity.

  O 1 O 2 O 3 O 4 O 5

- The AI project's goals and methods are not designed with a destructive purpose, but rather with safety, security, and of benefit to as many people as possible.

  O 1 O 2 O 3 O 4 O 5

- The AI project's goals and methods are aligned with fair and honorable business practices in general.

  O 1 O 2 O 3 O 4 O 5

- All those who design, build, and use this AI project have a practical stake in its use, misuse, and results.

  O 1 O 2 O 3 O 4 O 5

- The AI project protects the rights of the humans who generated the data used by it.

  O 1 O 2 O 3 O 4 O 5

- The AI project will utilize only those third-party tools and development partners who abide by the same high standards of explainable and ethical use of AI.

  O 1 O 2 O 3 O 4 O 5

If you wish, you can average the scores, but your project's ethical nature should become obvious from your answers.

AI Legal Questions

On a scale of 1 (unknown or not at all descriptive) to 5 (very descriptive), please rate the following characteristics of your organization and of the AI project you outlined in Section Three:

- The data to be used in my AI project is GDPR-compliant.

  O 1 O 2 O 3 O 4 O 5

- The data to be used in my AI project is compliant with HIPAA and/or other US data privacy laws, as well as comparable state laws or regulations.

  O 1 O 2 O 3 O 4 O 5

- My AI project will only use personal identification or behavioral data with explicit permission (or by making such sources anonymous).

  O 1 O 2 O 3 O 4 O 5

- My AI project poses little or no risk to individual rights, as defined by pending EU law or by the proposed AI Bill of Rights.

  O 1 O 2 O 3 O 4 O 5

- My AI project will not expose my organization to undue risk of legal liability.

  O 1 O 2 O 3 O 4 O 5

If you wish, you can average the scores, but your project's potential legal exposure should become obvious from your answers.

Equity and Inclusion Exercise

Use this worksheet to describe the equity and inclusion-related factors of the project you outlined in Section Three, at each stage of the AI development process:

AI Process Stage	AI Leaders/Stakeholders	AI Developers/Builders
Identify the use case or problem	How does the problem affect different demographic groups?	Design input from underrepresented or protected groups?
	Diversity in hiring practices and resulting team composition?	Purpose of the AI model documented (including diversity and sustainability factors)?
Model design, data collection, testing, & model development	In designing the AI model, with whom do we partner demographically?	How do we ensure diversity in the training data?
	With whom do we partner on data collection methods and sources, and in testing group diversity?	What are our best practices for ensuring high standards of ethics, transparency, privacy, etc.?

Deployment and monitoring	How do we build awareness on AI's impact (good and bad) on diverse demographic groups?	What are our accountability mechanisms for ensuring equitable and inclusive AI practices?

SECTION FOUR

Finding and Aligning the Right People

Section Two, on data readiness, included some of the personnel issues involved in a successful AI strategy. Chief among these issues is well informed, goal-aligned buy-in at the executive level. It also covered the organization's general data literacy and reality-based attitude towards artificial intelligence.

It is also important to know who else to include in your AI initiatives, especially those with the right technical expertise. This section will help you identify both the technical and the non-technical qualities of the people you will need.

PLAYBOOK

Popular fears regarding AI include the notion that it will radically change the work environment, displacing larger and larger portions of the human workforce. The fact is, AI-induced workforce reductions are likely, as they are with any wave of routine task automation. However, it is *more* likely that AI will spark the creation of many new jobs. It will also change the requirements for many existing jobs, not to take them away from humans but to make those humans more effective.

Not all these new or modified jobs will be for data scientists or other technical specialists. Many will be for those who know how to do their old jobs with new tools. Still, others will be for those who are needed to imagine what AI and data can do for their organization and put those systems to good use.

Who Do You Need?

Companies often announce technology-based initiatives without having all the necessary capabilities or personnel to complete them, or to integrate them into their existing infrastructure. According to a recent McKinsey survey,[1] only 5 percent of respondents said their organizations already had the capabilities needed to do so. When it comes to AI and data-centric initiatives, organizations must build the right institutional capabilities—not only sound technology and processes, as outlined elsewhere, but also the right *people* to make them work effectively.

In Section Three, we covered the overall human component of developing and maintaining valid AI capabilities. These include *executive engagement*—C-level leadership that understands the nature and purpose of AI and are capable of supporting projects appropriate to their business type. Executive buy-in will keep an AI team focused on the business problem they are trying to solve.

Another essential human element is the company's overall data literacy, especially for the non-technical members of the AI team. People should understand, as much as possible, the nature and value of data, and how it is being used to help make good

---

1 Guggenberger, Patrick, et al. "The State of Organizations 2023: Ten Shifts Transforming Organizations." *McKinsey & Co., 26 Apr. 2023,* www.mckinsey.com/capabilities/people-and-organizational-performance/our-insights/the-state-of-organizations-2023.

business decisions. They should also be able to clearly separate AI hyperbole and myth from its practical reality. Non-technical team members (or prospective team members) should always be encouraged to imagine why an AI solution would be beneficial, and learn to communicate that understanding to the data specialists who know how it can be done. As a company implements and scales its AI solutions, the results should be made widely known throughout the organization (within the confines of IP norms) to further advance overall data literacy.

The Data Team

Within any company reliant on efficient use of data, there will be individuals tasked with maintaining and improving it. As discussed in Section Three, IT departments may need to change the *status quo,* making data more accessible and less siloed, but the skill sets needed are fairly well defined. One thing is certain, AI will forever change the org chart of most organizations.[2] In no particular order, the team member roles can be described as follows:

> ***Chief Artificial Intelligence Officer***—The official title of "CAIO" is by no means mandatory, especially since AI can be new to an organization. The role may, of course, be assumed by a C-level executive already tasked with oversight of data, analytics, information management, or other related spheres. Responsibilities of this role may be handled at the managerial level, or even outsourced to a qualified outside consultant. Such a person is responsible for establishing the mission and purpose of the organization's AI strategy, coordinating implementation efforts, and working alongside those responsible for projects and their ethical and economic outcomes.

---

2 Raber, Zachary. "AI: Coming to an Org Chart Near You." *Medium, 15 Sept. 2023,* https://medium.com/@raber/ai-coming-to-an-org-chart-near-you-3581a59d2692.

AI Implementation Strategist or Director—This may be a combined role or split into two, depending on the size of the organization and the scope of AI initiatives. Another job title for this role could be that of a technical program manager, ideally someone with working knowledge of basic AI principles *and* the organization's business needs and goals. Such a person or persons would be responsible to work with the "CAIO" equivalent to identify opportunities and use cases where AI could benefit the organization in ways described in Section One. They must explore new AI opportunities, establish clear goals and performance metrics, recruit talent for the AI team, and work alongside engineers and content creators (within and outside the team) to keep projects on track and scale them when appropriate.

Data, AI, and Machine Learning Engineer—This is someone who understands the technical aspects of data, metadata, and the coding required to develop AI and ML models—either from scratch or as modifications to existing models. They may lack the desire or ability to push the boundaries of AI technologies or conduct research, but they should be able to implement project components in an agile and well-documented fashion.

AI or Machine Learning Analyst or Researcher—This is usually someone with the academic background required to push the boundaries of traditional IT or AI norms. Their charter is to discover and evaluate novel technical solutions that utilize big data and AI to solve business problems. Of course, they must be able to communicate meaningfully with data engineers, whether or not they themselves have the aptitude or patience to write code.

Data Scientist—Larger teams and projects will often require someone with the background and ability to obtain meaningful insights from data. They must use scripts and mathematical techniques to find solutions to specific problems, and typically have advanced academic and technical credentials—often at the PhD level—to stay current with contemporary research and to rigorously design data science and AI models, tools, and methods.

Of course, these roles, titles, and responsibilities will overlap to some degree. In the early stages, such roles may need to be outsourced, given the high demand for AI-specific talent. Also keep in mind that AI teams need not be strictly hierarchical and should always foster collaboration. A data specialist, no matter how brilliant, should never work in isolation. And, as stated earlier, the team must always be cognizant of the project's clear business goal, not collect or use data and AI for purely hypothetical reasons.

The order in which these roles are utilized or hired depends on the type of AI project you choose—and that fits your type of organization. For example, if you want to obtain meaningful insights from your internal data (a common mandate for Optimizer and some Extender organizations), you may want to identify or hire first a head of data who reports to your CTO or equivalent. Following that, identifying or hiring two or more data engineers and/or scientists, and give them time to understand the data, how they are obtained, and how to leverage the data most effectively.

Another common example is when you want to implement third-party APIs to achieve your AI goals, as detailed below. This generally means identifying or hiring data or machine learning engineers first, who would report to an existing CTO or IT manager for smaller projects—leading up to hiring a "CAIO" equivalent for larger, scaled AI initiatives.

Another example, more common with Innovator or Multiplier organizations, is when you plan to create *custom* AI models to achieve your business goals. The first hire in this case would be a data scientist or equivalent AI leader, or identify such a person already in the organization. Reporting to the organization's C-level leadership, they would, in turn, help find or hire the right people to implement their business data strategy.

In the workbook part of this section, you will explore ways to evaluate the existing human resources within your organization. You will also explore some of the criteria for identifying and hiring the right people to implement AI strategies with greater effect.

WORKBOOK

AI talent development has long preceded this year's explosion of interest in the topic. For many years, colleges and less formal "bootcamp" programs have been equipping coders with the practical means of sharpening their AI skills. Efforts to similarly equip business decision makers are also ongoing. (My book, *The AI Factor,* is among these.)

When considering how to find people with the right technical skills, always consider the following sources:

- ***Colleges and Universities***—In the past, proximity to a major university or technical school was considered an ideal way to recruit talent, by attending or sponsoring events, and by partnering with the institutions on various projects. Today, the geographic limitations are far less, but the financial support requirements limit the reach of smaller organizations. Nevertheless, according to US News and World Report,[3] there are a large number of academic institutions worth considering as having recruitment potential:

3 "2024 Best Undergraduate Artificial Intelligence Programs." *US News and World Report,* www.usnews.com/best-colleges/rankings/computer-science/artificial-intelligence. Accessed 10 Dec. 2023.

Rank	University	City
1	Carnegie Mellon University	Pittsburgh, PA
2	Stanford University	Stanford, CA
3	Massachusetts Institute of Technology	Cambridge, MA
4	University of California, Berkeley	Berkeley, CA
5	Georgia Institute of Technology	Atlanta, GA
6	University of Illinois Urbana-Champaign	Champaign, IL
7	University of Washington	Seattle, WA
8	Cornell University	Ithaca, NY
9	California Institute of Technology	Pasadena, CA
10	University of Texas at Austin	Austin, TX
11	University of California, San Diego	La Jolla, CA
12	University of Michigan, Ann Arbor	Ann Arbor, MI

- ***Recruiters***—For finding both technical and managerial members of AI teams, professional recruiters may be well worth the cost of their fees, which can range between ten and 45 percent of the first year's salary. Later in the workbook, you will find some criteria for screening and qualifying candidates.
- ***Job Boards***—These tend to have a wider reach than traditional recruiting agencies, but they also provide a larger pool of candidates to qualify, also using the same screening criteria.
- ***Networking***—This can take many forms, from attending conferences and meetups to referrals from existing investors and partners. In theory, this can provide a more qualified field of candidates, especially if the referral comes from a source that knows your business. The same screening criteria still applies, however.

The best practices for hiring, in general, are applicable to positions in an AI team. One of these is diversity—not only as a philosophical ideal but also because of the nature of AI itself. A diverse AI team is more likely to think broadly when choosing project priorities and goals—and to detect bias in the training data set. Other things to emphasize are common skill sets, such as the use of the same programming language, and prioritizing candidates who are adaptable problem solvers.

Existing Team Evaluation Questions

The first place to start is, of course, with your current management and technical teams. Without naming individuals, give an honest assessment of your team's current strengths. (Be sure to include yourself, where appropriate. Also, if such individuals are working with you on a contract or consulting basis, and not direct employees, include them in your evaluations.)

On a scale of 1 (unknown or not at all descriptive) to 5 (very descriptive), please rate the following characteristics of your AI team. If an individual described is not currently part of the team, rate the statement as 1.

- Our AI and data initiatives have buy-in from our executive leaders.

  O 1 O 2 O 3 O 4 O 5

- Our executive and managerial leaders understand the practical purposes of AI, and are not unduly influenced by popular misconceptions.

  O 1 O 2 O 3 O 4 O 5

- Our AI and data teams include a person whose role is that of a Chief Artificial Intelligence Officer, as described above.

  O 1 O 2 O 3 O 4 O 5

- Our AI and data teams include a person whose role is that of an AI implementation strategist, director, or technical program manager.

  ○ 1 ○ 2 ○ 3 ○ 4 ○ 5

- Our AI and data teams include one or more engineers who understand the technical aspects of data, metadata, and the coding required to develop AI and ML models, and have the coding skills required to implement AI projects effectively.

  ○ 1 ○ 2 ○ 3 ○ 4 ○ 5

- Our AI and data teams include one or more AI or machine learning analysts or researchers, as described above.

  ○ 1 ○ 2 ○ 3 ○ 4 ○ 5

- Our AI and data teams include at least one data scientist with the background and ability to obtain meaningful insights from data related to AI projects and initiatives.

  ○ 1 ○ 2 ○ 3 ○ 4 ○ 5

Low scores in these areas should not be cause for alarm. They simply mean that you need to enable your existing people to up their game, increase recruiting to complete your AI team, or possibly consider third-party partnerships.

Recruitment and HR-Related Questions

On a scale of 1 (unknown or not at all descriptive) to 5 (very descriptive), please rate your organization's recruitment and HR practices when it comes to your AI and data teams.

- Our organization has good, working relationships with academic and professional entities that facilitate our AI-related recruitment efforts.

  ○ 1 ○ 2 ○ 3 ○ 4 ○ 5

- Our organization is adept at using networking and external messaging to attract good candidates for AI-related positions.

  O 1 O 2 O 3 O 4 O 5

- Our organization makes good use of recruiters, job boards, and other sources of good candidates for AI-related positions.

  O 1 O 2 O 3 O 4 O 5

- Our recruiting and HR process is good at bias-free screening of CVs, conducting initial interviews, checking references, and otherwise insuring that candidates for AI-related positions are sufficiently qualified.

  O 1 O 2 O 3 O 4 O 5

- Our recruiting process for AI-related positions includes an adequate number of technical or strategic planning tests, free from undue bias, that indicate how well a prospective team member might perform.

  O 1 O 2 O 3 O 4 O 5

- Our HR process includes adequate means for evaluating AI-related job performance and positive team dynamics.

  O 1 O 2 O 3 O 4 O 5

Third-Party Partner Evaluation Questions

On a scale of 1 (unknown or not at all descriptive) to 5 (very descriptive), please rate the following characteristics of your organization and its AI partners. If an entity described is not currently part of your AI efforts, then rate the statement as 1.

- Our organization includes one or more technical project managers tasked with finding third-party AI developers, and then working strategically and practically with them.

  O 1 O 2 O 3 O 4 O 5

- We have a good working relationship with third-party developers capable of creating AI and data models suitable to our type of organization (Optimizer, Extender, Innovator, or Multiplier).

  O 1 O 2 O 3 O 4 O 5

- We have a practical understanding of third-party AI platforms, capabilities, and APIs, and how they may benefit our type of organization.

  O 1 O 2 O 3 O 4 O 5

- Our organization includes one or more engineers capable or implementing third-party APIs and performing other tasks necessary to implement third-party solutions.

  O 1 O 2 O 3 O 4 O 5

- Our organization has the capability to measure and evaluate the results of AI models and projects created by third-party developers, or by utilizing third-party APIs and platforms.

  O 1 O 2 O 3 O 4 O 5

- Our organization has been successful in utilizing third-party custom AI developers, or in utilizing third-party APIs and platforms.

  O 1 O 2 O 3 O 4 O 5

As with the internal team evaluations and HR practices, low scores simply indicate the need to evaluate the situation and plan an appropriate response.

Practice Exercises

No matter how your organization scores on these evaluations, it is never too late to think about how your existing AI team (and/or your AI development partners) should function optimally in pursuit of your business goals. Using the sample AI project you mapped out in Section One (or any other project, if you wish), describe the people or third-party partners involved in each step, their responsibilities, and any applicable notes or caveats. It is not necessary to identify specific persons or outside developers; simply describe them and their role in the process.

Step or Task	Person(s) Responsible	What They Must Do	Notes/Caveats
Identify project's business goals.			
Identify project's sustainability risks and benefits.			
Research project's data and technical requirements.			
Assign project's tasks, timelines, and measurement requirements.			
Collect, evaluate, and normalize the required data.			
Perform project's coding and testing steps.			
Conduct tests to measure project's accuracy and how it achieved goals.			
Report regularly on project results.			
Develop strategy to scale or adapt the project to new business goals.			

In doing this hypothetical (or real) exercise, you will likely see the need for more team members. So, to help sharpen your understanding of AI-specific recruitment processes, practice describing the steps involved. This falls into two broad categories:

- ***For leadership and managerial roles,*** focus on academic and professional credentials, as well as practical experience. Be aware of potential bias on your part, and of the economic, competitive, and creative advantages of diversity. Most of all, prioritize the skills and experience indicating a capacity for adaptable, creative, and collaborative problem solving.
- ***For technical roles,*** many of the above criteria still apply. However, the candidate's practical skills are paramount. Be ready to conduct real-world tests. These can include programming language tests, such as those available on platforms like Adaface (https://www.adaface.com/coding-tests) and elsewhere.
- ***AI-specific tests*** are also useful in helping identify those candidates best suited to your team. For example, you might show a candidate a set of forty or more images, some of which are identical, some close in to each other, some unique, and some with the same composition but of different subjects. A weak candidate will merely identify identical or similar images. A strong candidate will take this as a training data preparation problem, and will describe how the images might impact an AI model. A superior candidate would also be able to describe or create a means by which important image characteristics could be selected during the AI training process.

For each of the AI team roles named earlier in this section, describe their most important characteristics and skills, followed by what you, your recruiter, or your HR person might do to determine whom to bring on board. The point of this exercise is not to identify with particular individuals or types, but to summarize the things that make them a good choice for *your* organization.

Role	Characteristics/Skills	How To Discover This
Chief AI Officer (or equivalent)		
AI Implementation Strategist or Director		
AI/ML Engineer		
AI/ML Analyst or Researcher		
Data Scientist		

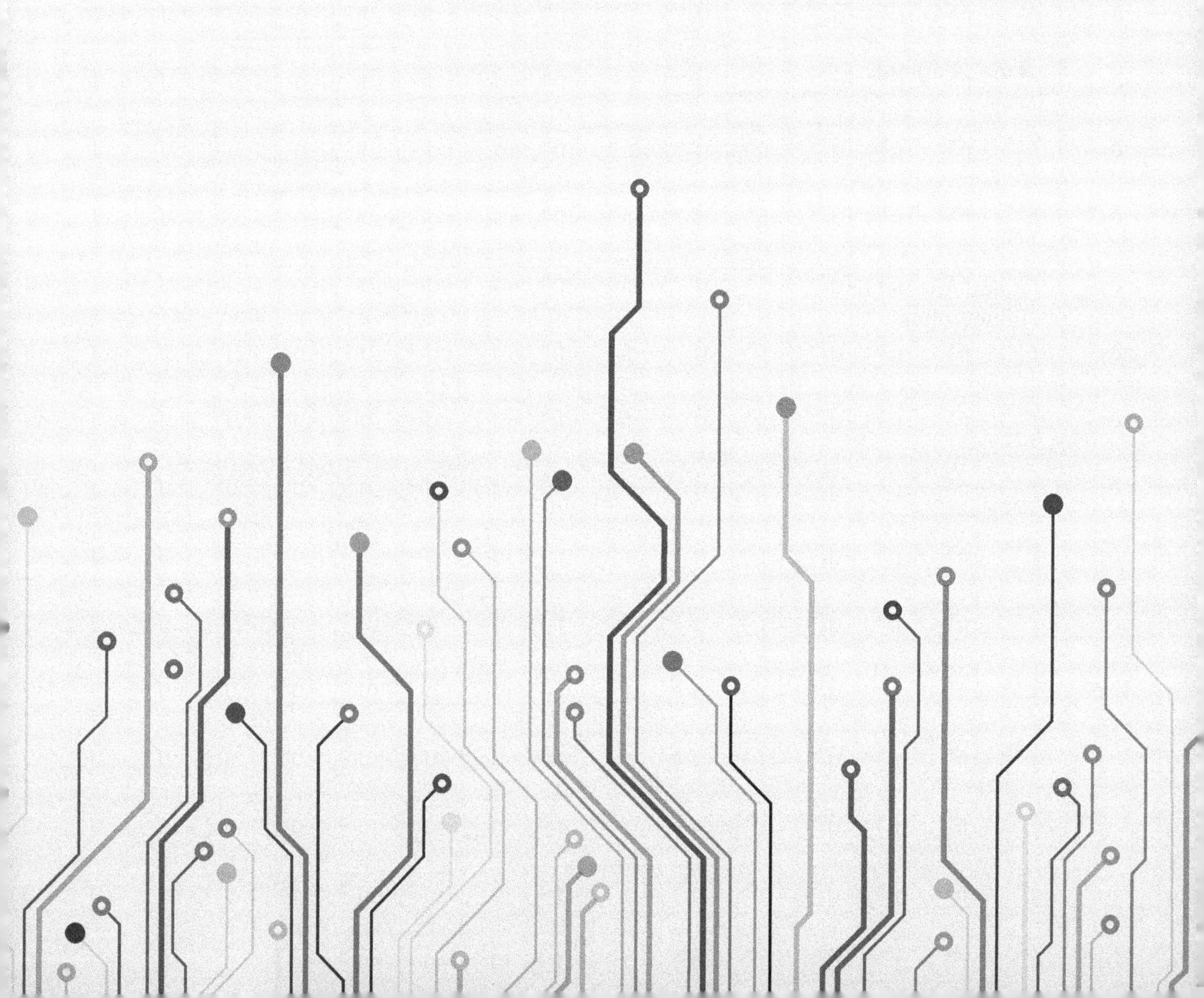

PART II

PUTTING AI TO WORK

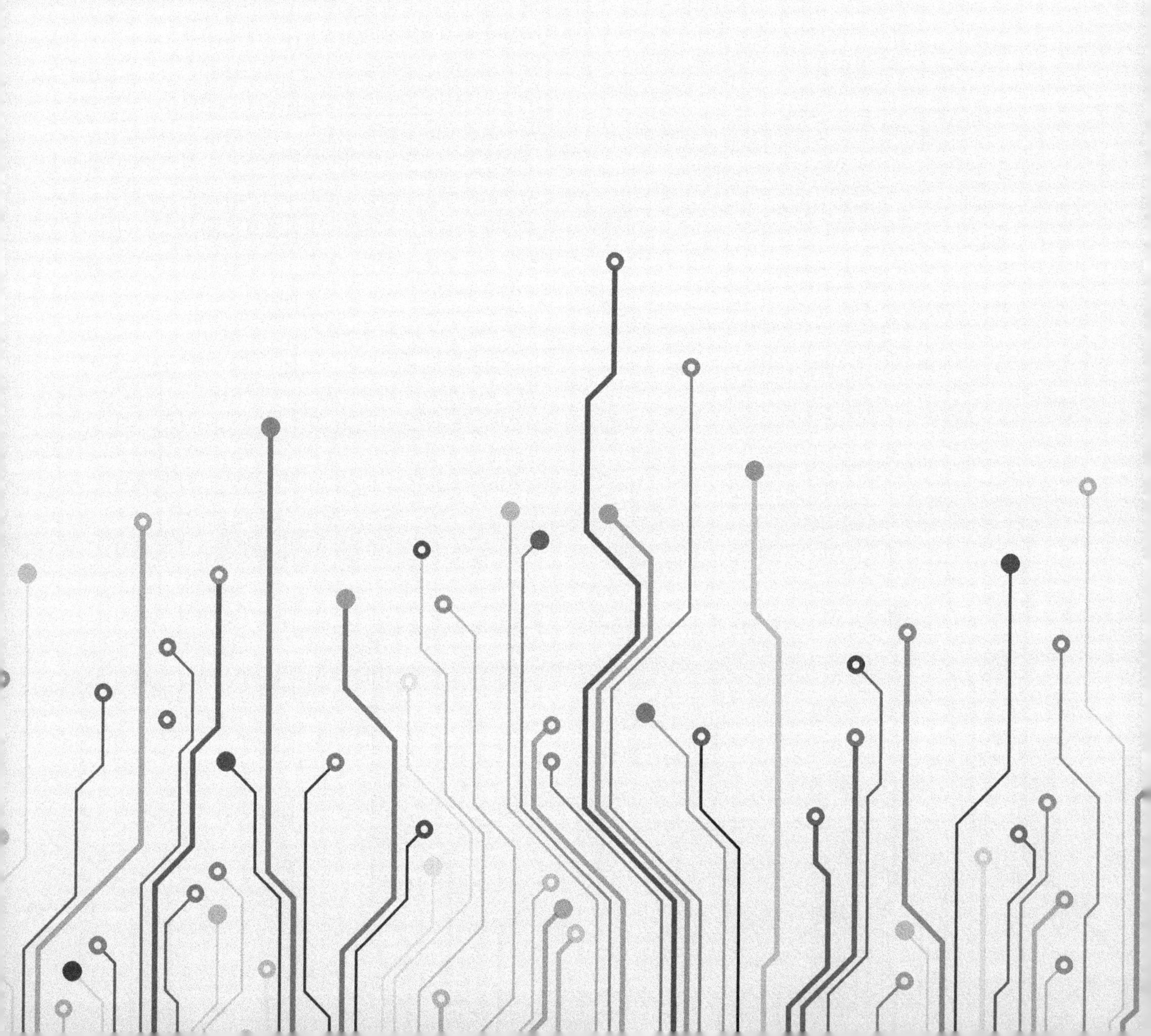

SECTION FIVE

Prioritization Techniques

The first four sections of this playbook are designed to lay the basic foundation for using data and artificial intelligence wisely. Once you have a grasp of basic AI strategy, sustainability, data readiness, and human resources, you will be better prepared to tackle the development and production of actual AI initiatives. The remaining sections will help you gain insights into the practical applications of AI in real-world scenarios.

Knowing your organization's nature and its data readiness state are essential, but your journey into AI and data-centric projects must start with a single first step. This section will help you identify and prioritize the AI use cases that offer the most value—and begin with the one that makes the most sense. It begins with evaluating the data value and the business value of a prospective AI project.

PLAYBOOK

AI Project Design—The Expanded View

In *The AI Factor,* I identified three general criteria for choosing a strategy for implementing artificial intelligence. The details of these criteria will be developed later in this book, but for now let's look at the big picture:

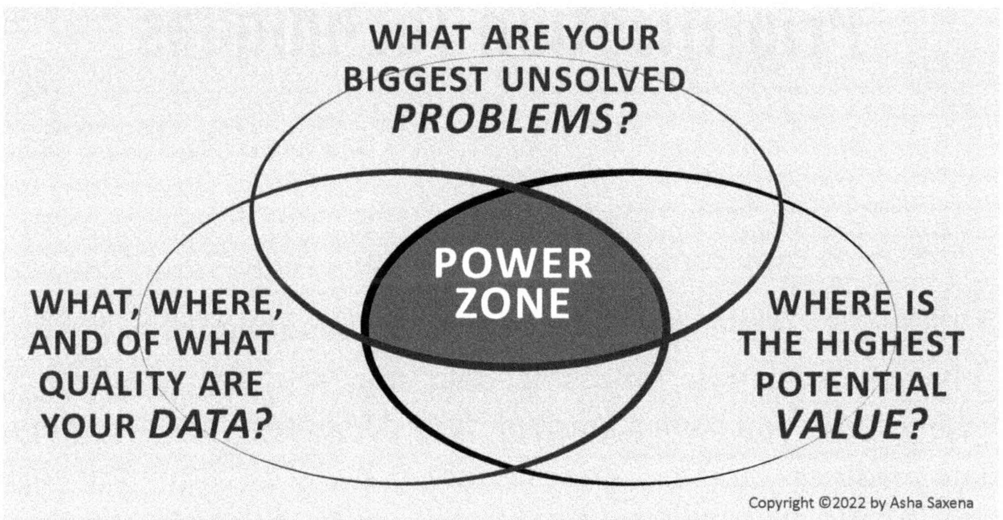

Take a moment to consider where your best AI strategy may lie. Here and in the workbook section, you should list as many potential projects as you can think of for your type of organization. The list can be very large, but it's important to set priorities. Once you have mastered an AI approach in one area, you can either scale it up, apply the same techniques to the next project, or preferably both. Sections Three and Four of this playbook will further detail the processes you need for analyzing your data readiness and assessing a data project's value, respectively.

The goal of any data project is to maximize efficiency, acquire new business capabilities, develop new processes (or products), acquire new customers and revenue, utilize new or better channels, or some combination of these. By listing, discussing, and prioritizing all potential AI projects to achieve these goals, an organization can come to a better understanding of itself, and begin to assess the requirements of putting their data to more effective use.

Focusing On the First AI Project

In the workbook section, we will discuss and start defining an AI project suitable for your type of business. It may be purely hypothetical, but ideally it should be a use of artificial intelligence that benefits your organization and, at least potentially, puts it on a path towards exponential growth.

In *The AI Factor,* I described several parameters that will help you identify the best possible AI project for your organization. The first task is to rate *the value of the data* you intend to use, by using the following scale:

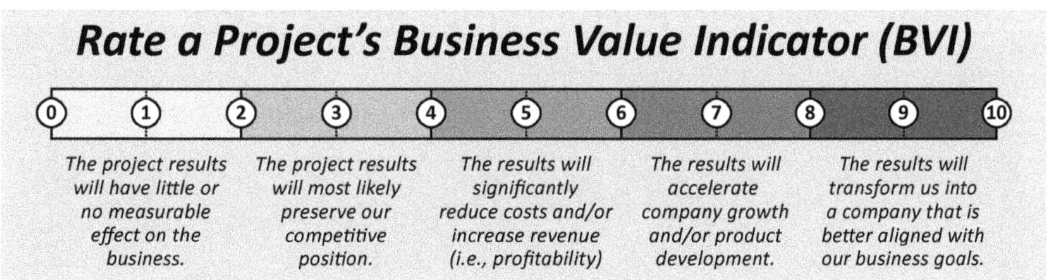

Obviously, a higher rating indicates that the project's underlying data are well suited to achieve your goals. This may require some conjecture on your part, but an initial rating will inform your decision on the viability of the project. The next step is to rate *the business value* of the project, using the following scale:

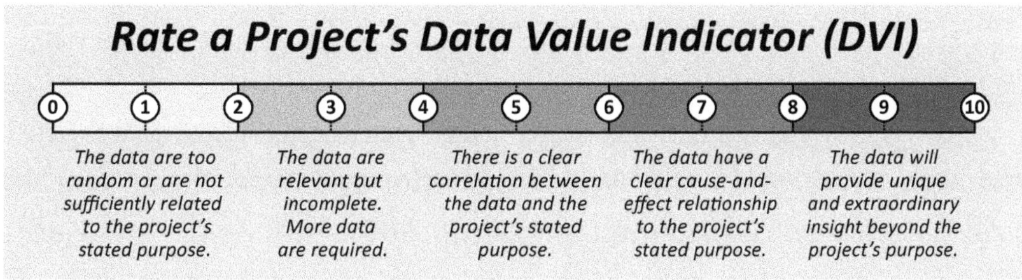

This too can be a subjective evaluation, so I strongly advise that you confer with others in your organization to counter any personal bias for or against a specific use of data and AI. Remember also that your first project is likely to be on a small scale, so guard against premature or inflated expectations.

When considering a project's business value, it is OK to consider what may happen if a successful project is implemented at scale. However, keep in mind that until your team develops experience at building and scaling AI projects, initial caution is highly recommended.

Finally, once you assign a rating to a project's underlying data and its business value, the last step is to combine those two variables. The result will indicate whether or not the project is within what I call its Data Performance Index.

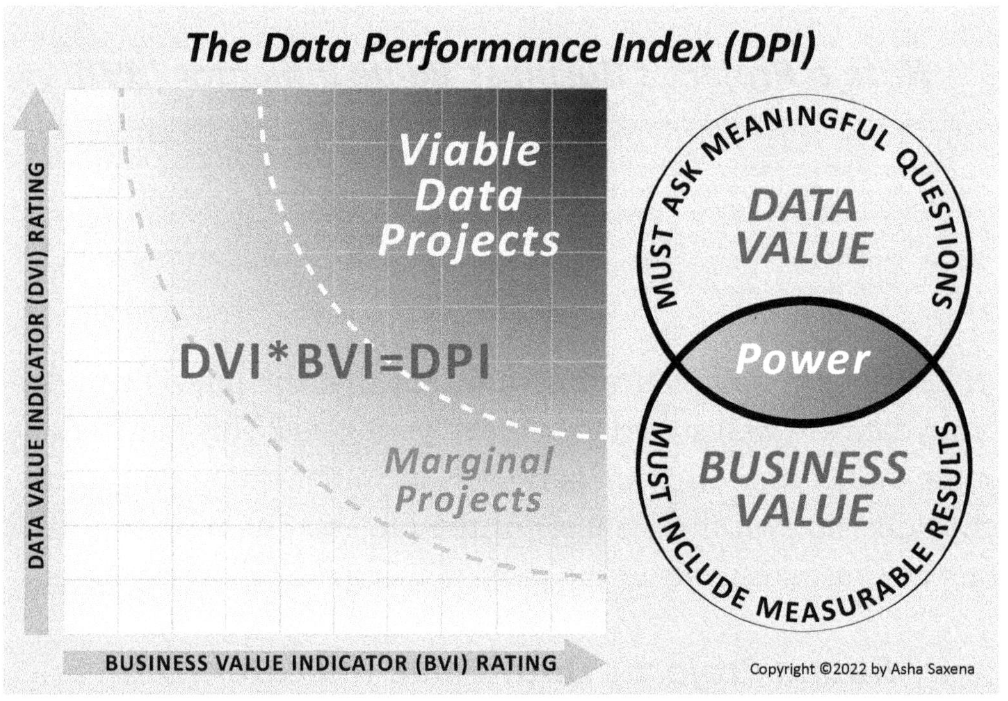

As you will learn through the workbook exercises—and through practice in real-world situations—rating these factors will depend on *asking meaningful questions* about the data and on *identifying the measurable results* that any data project would require.

WORKBOOK

Designing Your Own AI Project

Depending on the type of organization you belong to, the next step is to select a typical example and map out the particulars. It can be something from the sample project lists, or something else that suits your type of organization and its goals.

The following worksheet will help you identify the source(s) and type(s) of data you intend to use, give them a preliminary rating, and do the same for the project's business value. As you continue using this book, you'll be able to fill in additional details on sustainability, data readiness, and the process for setting up and running the project.

Workflow Diagram

In the space below, sketch a general, step-by-step workflow diagram of the project, including the anticipated actions or decisions that the AI project is expected to influence.

Data Description, Including Source(s) and Ownership

Brief Description of the Project's Business Goal

Data Value Questions

On a scale of 1 (not at all descriptive) to 10 (very descriptive), please rate the following characteristics of your proposed data set:

- All the data for the project are readily available, accessible, and secure.

  ○ 1 ○ 2 ○ 3 ○ 4 ○ 5 ○ 6 ○ 7 ○ 8 ○ 9 ○ 10

- The data are trustworthy, accurate, consistent, and reliable.

  ○ 1 ○ 2 ○ 3 ○ 4 ○ 5 ○ 6 ○ 7 ○ 8 ○ 9 ○ 10

- The data are from ethical, legally-compliant sources (e.g., GDPR).

  ○ 1 ○ 2 ○ 3 ○ 4 ○ 5 ○ 6 ○ 7 ○ 8 ○ 9 ○ 10

- The data are comprehensive, thorough, and clearly defined (metadata).

  ○ 1 ○ 2 ○ 3 ○ 4 ○ 5 ○ 6 ○ 7 ○ 8 ○ 9 ○ 10

- There is a clear correlation between the data and the project's stated purpose.

  ○ 1 ○ 2 ○ 3 ○ 4 ○ 5 ○ 6 ○ 7 ○ 8 ○ 9 ○ 10

- There is a measurable cause-and-effect relationship between the data and the project's stated purpose.

  ○ 1 ○ 2 ○ 3 ○ 4 ○ 5 ○ 6 ○ 7 ○ 8 ○ 9 ○ 10

- The data will provide clear insights related to the project's stated purpose.

  ○1 ○2 ○3 ○4 ○5 ○6 ○7 ○8 ○9 ○10

- The data will likely provide business insights beyond the project's stated purpose.

  ○1 ○2 ○3 ○4 ○5 ○6 ○7 ○8 ○9 ○10

Based on the answers, assign your project an *overall* data value score. This will be used to calculate the priority of your AI project.

Business Value Questions

On a scale of 1 (not at all descriptive) to 10 (very descriptive), please rate the following characteristics of your project's business goal:

- The project will achieve measurable cost savings for the organization.

  ○1 ○2 ○3 ○4 ○5 ○6 ○7 ○8 ○9 ○10

- The project will help preserve or improve our competitive position.

  ○1 ○2 ○3 ○4 ○5 ○6 ○7 ○8 ○9 ○10

- The project will help accelerate company growth and/or new product development.

  ○1 ○2 ○3 ○4 ○5 ○6 ○7 ○8 ○9 ○10

- The project will help transform the company into one more aligned with our aspirations.

  ○1 ○2 ○3 ○4 ○5 ○6 ○7 ○8 ○9 ○10

As you did previously, assign your project an overall business value score, based on the answers provided here. Then multiply the two to come up with your Data Performance Index or DPI. This will indicate the AI project's potential viability.

DVI * BVI = DPI

So, for example, if Project A's business value is 8, but its data value is only 2 (because the data sources were fragmented and incomplete), the resulting DVI of 16 would make it a marginal AI project at best. Similarly, if Project B had a high data value of 7 but a mediocre business value of 3, then the DVI of 21 would not justify significant investment. However, if Project C had both a high DVI and a high BVI, the results would make it a high priority project for your organization.

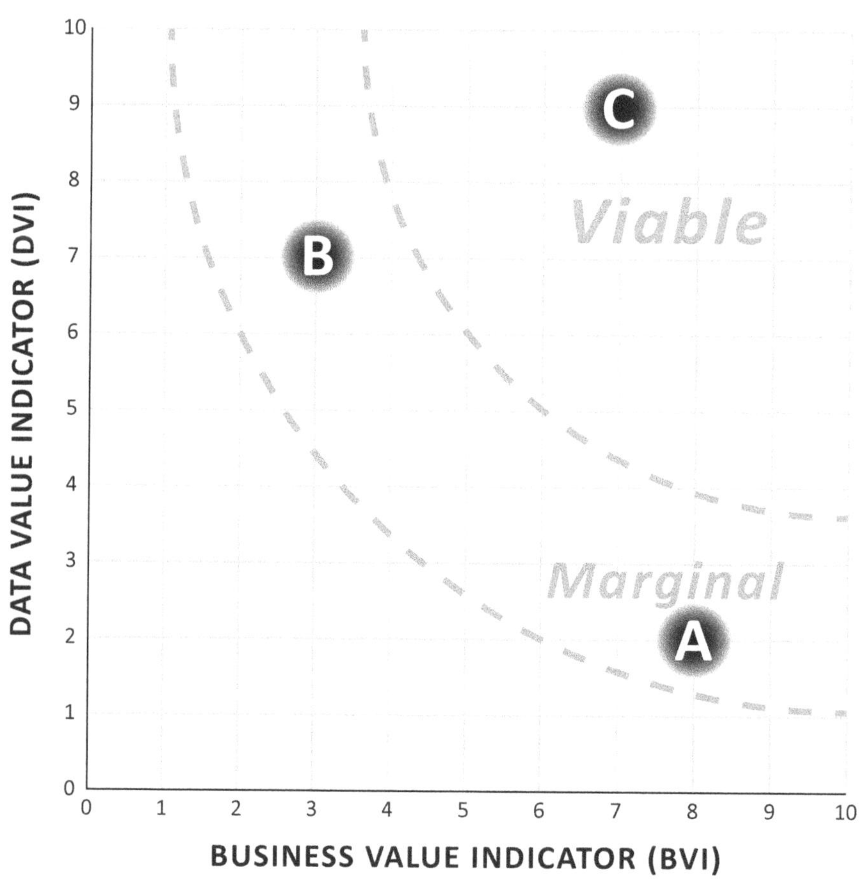

SELF-EVALUATION

Use the following chart to evaluate the DPI for your project.

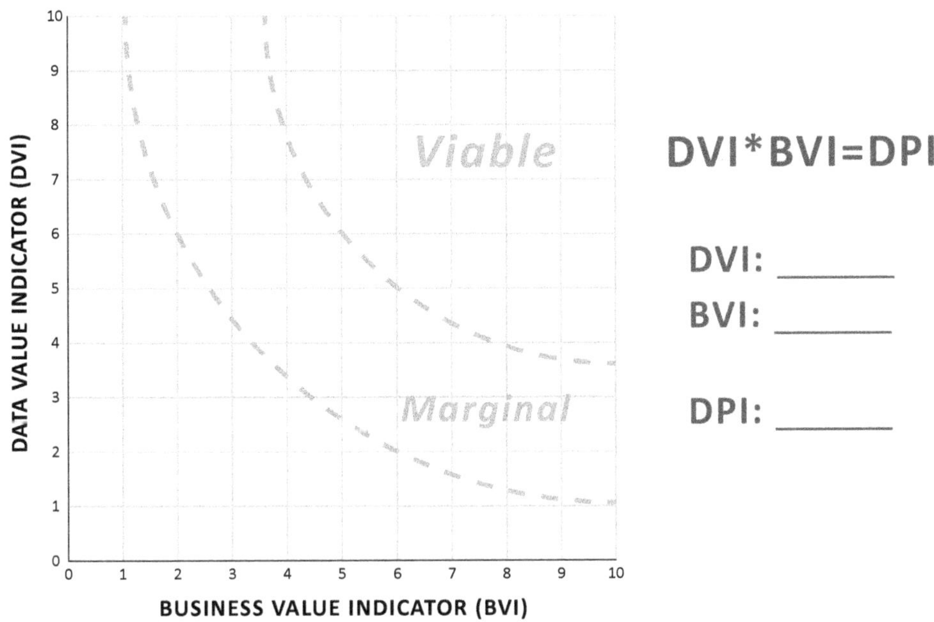

NOTES & QUESTIONS

Before making a decision on the project's priority, be sure to consider questions that may change its data or business value, and thus alter the final DPI. These may include questions such as:

- Can the quality or completeness of the data be improved?
- Can we afford to improve or augment the quality of the data?
- Can we discover ways to modify the projects parameters in ways that better align with our business goals (and thus be of greater business value)?

If these basic conditions cannot be altered, then the project's data performance is settled, and the project's priority level accepted.

SECTION SIX

Understanding AI Implementation

Once you have a general understanding of Data and AI as they pertain to your type of organization, and once you have mastered the general principles, it's time to consider specific implementation guidelines. These typically apply to any AI project.

PLAYBOOK

When planning the details of your AI project, think of your overall strategy as a big picture *canvas*. From this overview, you can develop each component in greater detail. (In the workbook part of this section, you will have the opportunity to create your own Data Power Canvas.)

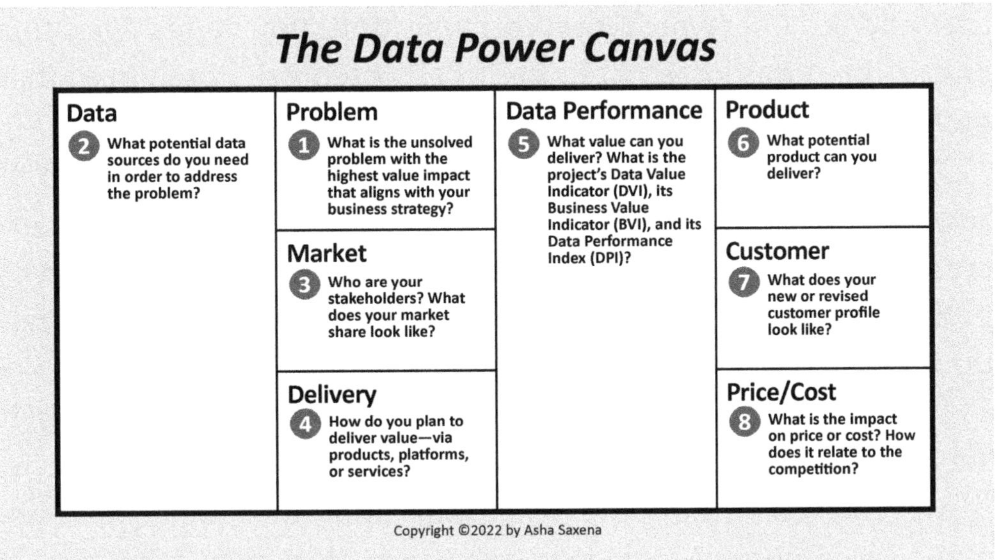

This is a simplified model, of course. Its purpose is not to set up a rigid, step-by-step protocol. Rather, its purpose is to define *all* the general parameters of a data-centric project, set realistic expectations for its potential, *measurable* impact, and to prepare for solving problems in the future.

Identifying the Most Pressing Problem

Every organization has multiple problems and barriers to success, each one vying for attention and resources. AI solutions can resolve many of these. But until your organization becomes adept at using data solutions at scale, it is wise to pick your targets one at a time. So, in order to pick ***the most important problem*** as candidate for an AI solution, always ask:

> **If this problem remains unresolved, then what does it cost us in terms of efficiency, revenue, or other tangible, measurable, and negative outcomes?**

As an experienced business leader, you will know instinctively your organization's most serious problems, but instinct alone is not enough. The secret is to pick ***one problem*** that is the most harmful and that can be resolved through the right implementation of data and AI. Other problems can wait until you've resolved this one. When you do, you'll be more experienced at solving other problems when the time comes.

There are several *types* of problems that AI is much better at solving than even the most experienced, intuitive business leaders.[1] These include business problems that require calculation of probabilities, correlations between multiple variables, and predictions based on "what-if" scenarios.

AI can certainly produce incorrect answers, especially if it acts without the domain knowledge of actual, human experts. However, even without overwhelming amounts of data, there are many cases where AI will lead to a good solution, if it is properly applied towards a well-defined problem. A few examples of such AI-solvable problems include:

General Business Problems—These can include inefficient or unresponsive customer relations, hiring and personnel issues, and the threats posed by fraud and other risk factors. Most of these problems stem from a lack of good information—or, more often, the inability to find that information in time to act on it in an efficient manner. *The AI Factor* describes many other such problems, but you can undoubtedly find more of them within your organization.

Problems Related to Market Conditions—These can be external, even global conditions over which we have no control. They can also be related to our own customers or prospects—what they want, what they actually buy, and where and when they do so. Without knowing these factors, organizations are very likely to fail, not only in the effective sales and marketing of existing products or services but also in their ability to develop new and innovative ones.

1 Suh, Bob. "When Should You Use AI to Solve Problems?" *Harvard Business Review, 17 Feb. 2021,* https://hbr.org/2021/02/when-should-you-use-ai-to-solve-problems.

Delivery or Fulfillment Problems—Every organization on earth exists to provide *something* of value to another organization or individual, hopefully to everyone's mutual benefit. There are countless barriers to doing so effectively, including equipment breakdowns, supply chain disruptions, poor communications, and just plain old human error. Of course, AI cannot eliminate these, but it can predict them with reasonable certainty, suggesting practical, even innovative ways to reduce their impact on an organization's ability to thrive.

In the workbook part of this section, you will have the opportunity to identify one such problem, judge its severity, and determine if a viable AI solution exists.

Selecting the Right Data

In Section Two, we covered the general principles of ***data readiness,*** not only on the practical issues of data management, usage, and literacy, but also on human and business factors such as executive buy-in and strategic alignment. To review, a viable AI project must have these general principles in place:

Good Data Management—Your data must be accessible, secure, and of sufficient quality to be used by artificial intelligence and machine learning algorithms with a reasonable level of confidence. Such data must also be sufficiently integrated, not siloed or unrelated to other data sources.

Informed Engagement—Your organization must have sufficient engagement by the human beings responsible for the use of data and AI. This begins at the top, with executives who understand and are responsible for the outcomes of AI initiatives. It must include all those who interact with the data, understand their part in its use, and are accountable for its application.

There are many potential sources of appropriate data for use in AI projects. Some of these are described throughout *The AI Factor*, with many others specific to your individual organization. These include external data sources, both public and private. But it is important to first consider data that you may already possess, including but not limited to:

Call center records from customer service or technical support. These can include text threads, recordings, and survey results. Even unstructured data from these sources can yield valuable behavioral information, including why your customers keep doing business with you or why they leave.

Sales, installation, or incident reports by salespeople, customer service reps, or by field service agents or trainers. Even simple transaction reports or demographic breakdowns can yield valuable information that AI can use.

Marketing funnel data from customer relationship, sales, and content management systems can yield not only valuable insights on customer behavior and preferences, but also evaluations of the marketing content itself.

Ecommerce records and reviews contain not only detailed historical transaction history but also valuable insights into product preferences and behavior. This applies to the purchasing and usage behavior of both consumers and businesses.

Social media and search results data are typically unstructured, but it can provide valuable information. This includes your organization's products, customers, and reputation, as well as of your competitors. These sources can reveal customer preferences, potential referrals, performance issues, and even new business opportunities.

In the workbook part of this section, you will identify the data sources relevant to your AI project, including an assessment of its quality and overall value. Remember that simply having tons of data does not guarantee a good outcome. An AI project must have an explainable, measurable goal, and the involvement of human experts accountable for the results.

Choosing the Right AI Approach

Artificial intelligence is not a mysterious black box, whose workings are known only to data scientists or engineers. As we covered in Section Two, viable AI must be explainable to competent, non-technical decision makers. Actual human beings must be able to measure and judge an AI system's success or failure—and be held accountable for both.

That said, there are several AI methods to consider when planning a project. While it requires the expertise of people described in Section Four to actually implement them, the methods themselves can be understood by non-technical team members. In no particular order, they include:

> ***Machine Learning*** (ML)—Perhaps the most common approach, this uses structured or unstructured training datasets, algorithms, and artificial neural networks to improve its results over time. It uses the training data to mimic the human learning process. In fact, ML (and its more advanced iteration, deep learning) is literally the foundation of all forms of artificial intelligence, as described in *The AI Factor*.
>
> One facet of ML, ***supervised learning,*** involves training AI models by using labeled data, often with a human agency to validate the results. Once trained, the AI will be able to predict or respond to similar situations with increasing accuracy. Another level, ***unsupervised learning,*** is used where the data are unlabeled or unstructured. Its goal is to detect patterns, relationships, structures, and anomalies, especially where those gathering the data may not be aware of such patterns.

A third ML approach, ***reinforcement learning,*** allows the AI model to learn by trial and error, with humans affirming positive results and tagging negative ones. This is how AI systems can become more proficient at skills such as chess or Go, by playing against human experts. It is far more problematic in areas of higher risk, such as autonomous vehicles.

Computer Vision (CV)—This is an often misunderstood and misapplied AI approach, involving the processing of visual information. It also involves the use of a training data set, extracting information from still or video images, and using that data to detect patterns, recommend actions by humans, or take action automatically. There are several aspects of CV, including image recognition, object detection, image restoration, and image segmentation. A common application is the facial recognition AI models used by government law enforcement and private security entities. These can carry a high degree of risk, due to bias in the training data and unresolved privacy issues.

A narrower, less controversial application of CV is biometric recognition. AI-enabled security systems can recognize individual retinas, fingerprints, and faces to prevent unauthorized access sensitive locations or systems. Still less controversial, but not without risk, is the use of CV in medical imaging. Properly trained, such systems can use AI pattern recognition to assist medical professionals in diagnosing cancer and other conditions.

Computer vision relies on two factors, ***sensitivity and resolution,*** to accurately recognize meaningful patterns within image data. Both factors affect the CV system's ability to identify small details, reducing the likelihood of false positives. They are also essential in high-risk situations, such as inspection of infrastructure, manufacturing quality control, and scientific analysis.

Natural Language Processing (NLP)—This is by far the most prominent and misunderstood AI methodology, thanks to the explosion of interest and/or hysteria over ChatGPT and the like. Its foundation is known as a large language model, or LLM, using vast amounts of text data, logical text models, and closely-held algorithms to mimic human responses to text queries or prompts.

NLP uses several techniques or steps to mimic human conversation and predict the words that should appear next in a sentence or paragraph. One of these is ***text preprocessing,*** translating raw text into computer-readable formats and extracting relevant information in the process. The next is ***part-of-speech tagging,*** defining the meaning of a word based on its context. This is especially useful, but not infallible, with complex and confusing languages like English.

Other aspects of NLP are ***named entity recognition*** and ***sentiment analysis.*** The former is used often to recognize, extract, and identify specific persons, places, or events within large volumes of text—an essential component of modern search engines. The latter can be used to determine the emotional context of the writing, whether it is positive, negative, or neutral, for example. This can be done broadly, to evaluate the overall mood of many writers on social media, for example. It can also be used on an individual basis, to detect increased stress in a chatbot user and hopefully use that to switch the conversation to a human.

Today's NLP models come with inherent limitations and uncertainties. OpenAI founder Sam Altman and others have cautioned that the growth potential of LLMs is not unlimited. The latest iteration of ChatGPT relies on *trillions* of variables (up from 175 billion with GPT-3). Even so, such systems are prone to AI "hallucinations"—results that seem plausible but are partially or completely inaccurate. Simply adding more data is unlikely to resolve the issue, even if one assumes the data supply is unlimited. (Privacy concerns and the advent of Web3 technologies to protect private data would suggest that there will be limits on LLM mining of "unlimited" data.)

Automation and Robotics—These are not AI techniques per se, since they can exist without artificial intelligence or machine learning. However, the addition of AI and ML to automated or autonomous systems has increasingly put them in the realm of possible approaches to using AI.

Creating Measurable Results

The right-hand portion of the "Data Power Canvas" is crucial to any AI project. There must always be a clearly-defined business goal—the counterpart to the pressing business problem defined earlier. Without a reliable means to measure its effectiveness, an AI initiative is just a cool science project.

There are three broad categories to consider at this point. One relates to the ***product*** (or service) to be created or improved through the application of AI. This need not be a wholly new, breakthrough product like the first iPhone or a smash hit on Netflix. Such products are typical to multiplier organizations, as discussed in Section One. An equally valid product for an innovator, or even an extender or optimizer organization might be a meaningful change to an existing product or service. The point is that the resulting product or service must make a meaningful, usually financial difference to the organization. Measuring that difference is the means by which an AI model can be evaluated.

The second category of measurable AI goals involves the ***customer***—or more specifically, the behavior of a group of customers towards the organization. This can include sales metrics, net present value, reputation and referrals, and a host of well-known measurements of success or failure. This category is common to all four types of organizations, since all businesses desire financial growth. Even nonprofit entities can fit in this category, since they all seek a meaningful level of connection with a growing number of supporters.

The only caveat in this category is to be sure that the AI model being developed identify a *specific* aspect of measurable impact when it comes to customers. It's all well and good to aim for "a better relationship" with customers, but unless the AI project has a specific, measurable result, it cannot justify the project's cost.

The third category of AI goals, ***price and cost,*** is most often applicable to optimizer and extender organizations, but may apply to others. The measurement here is simple: Will the AI approach reduce costs or allow the organization to increase prices. So long as the goal is specific, it will readily prove (or disprove) the premise of the proposed AI model.

Mastering the Details

A valid AI strategy also requires a solid grasp of the practical realities. In addition to having organizational data readiness (Section Three) and the right people (Section Four), you must also pay attention to AI's operational requirements. These include the following:

- ***Hardware and Other Infrastructure***—This means not only sufficient storage and processing power (local or cloud-based) to handle the calculations. It also means having the means to assess current hardware performance and review ongoing hardware and infrastructure requirements.
- ***Programming Language Standardization***—Technical team members and partners must use common programming standards allowing for optimum simplicity, speed, and ease of documentation and review.

- ***Techniques***—All team members must adopt a common set of techniques appropriate to their problem domain. These include issues of concept generation, assignment, and forecasting. The team's techniques must allow members to validate or challenge characteristics of the data. They must provide the best possible combination of accuracy, development speed, runtime efficiency, and results transparency.

Executing an AI project is no small task, even when it has a well-defined, measurable business goal. But with the right people and the right data, the use of AI can transform your organization.

WORKBOOK

Implementing a strategically relevant AI project is a complex, detailed process, requiring specific skill sets and personnel, as described in previous sections of this book. By going through the following questions and exercises, you will become better able to implement your AI project wisely.

Basic Business Questions

To help you refine your project, consider the following business questions. On a scale of 1 (unknown or not at all descriptive) to 5 (very descriptive), please rate your organization on the following issues. (This is not an exhaustive list.)

- Our customer relations process is efficient and responsive, adding tangible value to our reputation and profitability.

  O 1 O 2 O 3 O 4 O 5

- Our hiring and personnel management processes are both efficient and fair.

  O 1 O 2 O 3 O 4 O 5

. Our organization and our customers are well protected against fraud.

O 1 O 2 O 3 O 4 O 5

. Our sales and marketing process is efficient—focusing on the known needs of prospects and existing customers—and able to effectively close business and satisfy customers.

O 1 O 2 O 3 O 4 O 5

. Our organization is providing the right products and services to a well-defined and sufficiently large customer base.

O 1 O 2 O 3 O 4 O 5

. Our organization is efficient when it comes to delivery or implementation of our products or services.

O 1 O 2 O 3 O 4 O 5

. Our organization is well aware of potential opportunities to offer our products or services to new customers, and is prepared to do so.

O 1 O 2 O 3 O 4 O 5

. Our organization is able to develop new products or services, and to identify customer demand for those products or services.

O 1 O 2 O 3 O 4 O 5

. Our organization is prepared to create strategic partnerships or acquire other assets or, organizations.

O 1 O 2 O 3 O 4 O 5

. Our organization is able to adjust quickly to changing local or global market conditions.

O 1 O 2 O 3 O 4 O 5

There is no need for an average score in this section. Simply discuss and identify the most pressing problem—i.e., the one with the *lowest score*—as the most likely candidate for an AI solution. Keep in mind that such a candidate should always be in line with your *type* of organization, as defined in Section One, and the availability and useability of relevant data.

Selecting the Right Data

In Section Two, we reviewed the general requirements for data readiness, which include strategic objectives, good data management practices, and your company's capacity to utilize data effectively. Here, you will zero in on the specific data you'll need in order to implement the AI project that you've identified.

On a scale of 1 (unknown or not at all descriptive) to 5 (very descriptive), please rate the following characteristics of the data to be used in this AI project:

- The data are relevant to the business, market, or delivery problem to be addressed by this AI project.

  o 1 o 2 o 3 o 4 o 5

- The data will have a direct, positive impact on the organization's products, customers, or the price or cost of a product or service.

  o 1 o 2 o 3 o 4 o 5

- The source of the data is fully known, and there are no "missing pieces" where data are concerned.

  o 1 o 2 o 3 o 4 o 5

- The data are readily accessible and not siloed or in multiple, unrelated databases.

  o 1 o 2 o 3 o 4 o 5

- The data are owned by the organization and/or the data are publicly owned and/or the data are from sources from whom the organization has obtained permission to use.

  O 1 O 2 O 3 O 4 O 5

- The data are stored securely, in the cloud or locally, and can be accessed only by authorized personnel.

  O 1 O 2 O 3 O 4 O 5

- The data are highly structured, with consistent, logical metadata and of sufficiently high quality to be used reliably in data-driven solutions.

  O 1 O 2 O 3 O 4 O 5

- The data are unstructured but of sufficiently high quantity and quality to allow for AI and machine learning detection of meaningful patterns.

  O 1 O 2 O 3 O 4 O 5

- The data have previously been used successfully in developing AI solutions to business problems.

  O 1 O 2 O 3 O 4 O 5

- The organization has historical data that can be used to test the validity of AI projects.

  O 1 O 2 O 3 O 4 O 5

Mapping the Project

Using the information and insights gained from these questions (and from your findings in Section Five), create your own Data Power Canvas for the project.

The Data Power Canvas

<table>
<tr><td rowspan="3">Data</td><td>Problem</td><td rowspan="3">Date Performance</td><td>Product</td></tr>
<tr><td>Market</td><td>Customer</td></tr>
<tr><td>Delivery</td><td>Price/Cost</td></tr>
</table>

SECTION SEVEN

Scalability (the AI Flywheel)

A single AI project, even one that exactly suits your type of organization, is unlikely have a major impact. More will be needed in order to make AI and data-centric initiatives broadly beneficial to your business. This section will explore how to implement, measure, and scale AI initiatives more effectively.

PLAYBOOK

In 2001, business author Jim Collins popularized the use of the flywheel concept to describe how organizations needed to act in order to achieve breakthroughs and outdistance the competition.[1] As with a literal flywheel, the initial effort may seem to have minimal results. But the secret to success, Collins writes, is in the cumulative effect of persistent effort, applied in a consistent direction over time. As it is with business process in general, so too is it with the application of artificial intelligence. A single

---

1 Collins, Jim. *Good to Great: Why Some Companies Make the Leap...and Others Don't. HarperBusiness, 2001.*

project may seem to have little overall impact on business, but if the project scales, or inspires new AI initiatives, the results can have a multiplier effect.

Simply stated, the "AI Flywheel" starts with a relatively small initial application, such as the one you've mapped in this workbook. If planned well, the project will generate new data and business insights, which will fuel further AI-driven enhancements—whether in the form of the same project applied more widely or in the form of entirely new projects. The increased momentum will accelerate over time in a self-reinforcing cycle leading to greater cost savings, competitive advantages, and business growth.

In *The AI Factor*, this concept was presented as a continuous cycle of six steps:

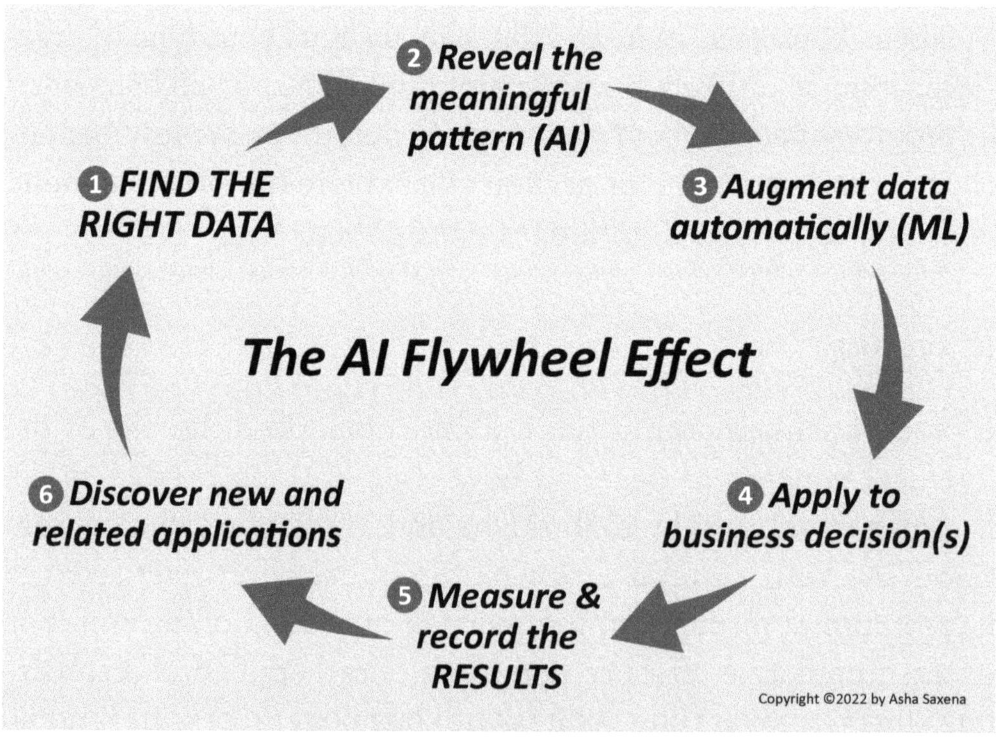

All AI projects must begin with the right data, as we've learned throughout this book. The first step, of course, if to make sure you

have access to the right data—of sufficient quantity and *quality* to reveal meaningful patterns in the data. In many cases, the data will be augmented continuously through the machine learning process. This will further refine and detect meaningful patterns, which can then be applied to business decisions. As the results are measured and recorded, two things happen. The original project can be continuously improved ***and*** new and related applications can be discovered. Then, the data generated by either the original application or related ones become the source for new and greater projects.

The Benefits of Scaling AI

For organizations that have established their general data readiness, the flywheel effect can speed up ***and optimize*** the flow of information between humans and machines, creating a self-reinforcing mechanism for continuous improvement and innovation. In other words, by finding truly meaningful patterns in the overwhelming volume of information faced by business professionals today and condense it to more manageable levels.

Scaling AI capabilities can also give organizations greater ability to predict future outcomes, whether they be related to new business opportunities or threats to the company's success. By the same token, it can guide the company's efforts to recruit new talent and warn of possible unethical or illegal implications presented by a particular course of action.

Above all, the flywheel effect gives organizations a greater potential to succeed over the long term, by compounding overall business value of existing data, and by providing high-quality data for new AI initiatives.

Rules of Thumb for Scaling AI Capabilities

Scaling an organization's AI capabilities will require an increased commitment to the principles stated in Part I, including a firm understanding of your organization's type and its data readiness state. It also requires paying close attention to the sustainability, explainability, ethical, and legal aspects of AI, as well as the people needed to sustain it.

The new discipline for scaling AI initiatives has been termed Machine Learning Operations or MLOps.[2] It includes ***standardization*** of every step in the development process, from collecting and preparing the data to creating the model's features, to training and fine-tuning the model, to validating that it works. When a model is ready for deployment, every aspect must be carried out in a consistent, documented manner, from monitoring output and performance to ensuring that it is sound ethically and from a legal compliance standpoint.

A second requirement for effective scaling involves the people needed to develop increasingly ambitious AI projects. As explained in Section Four, this requires a multidisciplinary team capable of clear communication and mutual support.

Finally, scaling AI operations always requires the wide selection of tools. Like the people who use them, AI tools and libraries must be able to talk to one another. Interoperability is key, allowing the organization's tool set to support creativity, speed and safety. Highly-specialized AI code and tools, however attractive they may seem for a single project, may ultimately hinder an organization's ability to scale.

Closing the Gap

For many organizations, creating or expanding/scaling AI capabilities must include contracting with *outside* teams of specialized AI developer partners. Below are some partial lists of developers with AI capabilities and APIs that may become necessary to bridge the gap between your current team's abilities and the optimum

2 Vartak, Manasi. "How to Scale AI in Your Organization." *Harvard Business Review, 10 Nov. 2022,* https://hbr.org/2022/03/how-to-scale-ai-in-your-organization.

capabilities you will need. Of course, unless you are planning to outsource your entire AI project, there must be someone in your group capable of implementing third-party APIs. First up are four of the major AI players: Amazon, Google, IBM, and Microsoft.

All four companies offer APIs for their capabilities in handling ***speech and text,*** including their chatbot tool sets, intention analysis, key phrase extraction, language recognition, sentiment analysis, and speech recognition. All four also offer similar support for text-to-speech conversion, topic extraction, and translation. Microsoft has the highest number of AI-related APIs in this area, followed by IBM.

Not surprisingly, Google dominates in the area of ***image analysis*** APIs, followed closely by Microsoft. These APIs can enable AI projects requiring facial detection and analysis, as well as text and object detection. All four companies offer APIs involving inappropriate content detection.

Finally, when it comes to AI-related APIs involving ***video,*** Microsoft has the most to offer (14) followed by Amazon. As of this writing, Google offers very few, and IBM none. Microsoft, Amazon, and Google each provide APIs for detecting inappropriate content, objects, and scenes.

Using third-party AI resources is not limited to using the APIs of tech giants. There are vast stores of free, documented, open source code available to organizations, mostly accessible via GitHub (https://github.com/) and elsewhere. Of course, as with using tech companies' APIs, you must have the right people on board to take full advantage of the open source code's potential.

Finally, in between the tech giants' resources and those of the open source world, there is a fast growing network of developers who will develop AI capabilities for you. While not free of charge, these companies can fill many existing gaps in your organization's AI capabilities. They can also compensate for incomplete AI teams. A good AI implementation strategist or project manager can, for example, be the leader/coordinator of an outsource developer.

The following is a partial list of such resources, with some of the specialties each one can provide:

Custom AI Project Development

- Rchili (https://rchilli.com)
- Reconess (https://reconess.com/)

Forecasting (SaaS Platforms with APIs)

- Ayadi (https://www.ayasdi.com/platform/)
- Infosys Nia (https://www.edgeverve.com/artificial-intelligence/nia/)
- Unplugg (https://unplu.gg/test_api.html)

Language (SaaS Platforms with APIs)

- Aylien (https://aylien.com/)
- Automated Insights (https://automatedinsights.com/)
- Meaning Cloud (https://www.meaningcloud.com/)
- Spot Intelligence (https://www.spotintelligence.com/)
- Tisane (https://tisane.ai/)

Vision (SaaS Platforms with APIs)

- Clarifai (https://clarifai.com/developer/guide/)
- EveryPixel (https://api.everypixel.com/)

Finding the right people for your AI initiatives is challenging, given the high demand for and limited supply of the right talent. But whether you recruit that talent or find outsourcing alternatives, it is an absolute imperative.

WORKBOOK

In order to create the flywheel effect in your organization's AI initiatives, you will need to document the six steps outlined in the workbook. The following exercises will help you do that for the project you've outlined in Sections Five and Six.

The Right Data

In your opinion, on a scale of 1 (low confidence) to 5 (high confidence), rate the following characteristics of the data you intend to use:

- The data are of sufficient volume to warrant use in the AI project.

  O 1 O 2 O 3 O 4 O 5

- The data are well-structured (consistent metadata) or contain significant, discernable patterns that would allow for detection of meaningful patterns.

  O 1 O 2 O 3 O 4 O 5

- The data are easily accessible.

  O 1 O 2 O 3 O 4 O 5

- The data are not scattered or siloed in disparate, unconnected data repositories.

  O 1 O 2 O 3 O 4 O 5

- The data require little or no validation or reconciliation to make it suitable for use in the project.

  O 1 O 2 O 3 O 4 O 5

Next, write out your reasons for these ratings, and any other relevant comments, in the space below. If possible, indicate ways in which these ratings might be improved. After the project has been completed, go back to your original assessments and note whether your ratings were born out—and what can be done to improve them.

Data Notes (BEFORE)	Data Notes (AFTER)

Meaningful Patterns

In the space below, indicate the patterns you expect the AI to reveal, and their relevance to the business problem or other issue mapped out in Section Six. After the project has been completed, note whether or not such patterns were revealed.

Pattern Notes (BEFORE)	Pattern Notes (AFTER)

Data Augmentation

In most cases, your AI project will involve the data being augmented during the machine learning process. In the space below, indicate to what extent you expect this to happen, and what the nature of the augmented data will be. After running the project, note the actual results.

Data Augmentation Notes (BEFORE)	Data Augmentation Notes (AFTER)

Impact On Business Decisions

In Sections Five and Six, you will have evaluated the business value of your AI project. In the space below, briefly note both the anticipated and the actual application of the results to your business.

Business Application Notes (BEFORE)	Business Application Notes (AFTER)

Results

In the space below, briefly note the anticipated results you intend to measure and record—and the means or techniques you intend to employ. After running the project, indicate how well the results were measured and recorded, plus any insights gained during the process.

Results Measurement Notes (BEFORE)	Results Measurement Notes (AFTER)

Discovery

Based on the results of your AI project, successful or not, your team should always explore the project's implications for new or expanded AI applications. This, in turn, will point to, and in some cases actually create new data for the next project iteration. In the space below, note the possibilities for new or related AI applications implied in your project's results. After the project has been completed, note any new or unexpected discoveries, and whether they have generated data that will support new AI projects.

Discovery Notes (BEFORE)	Discovery Notes (AFTER)

Finally, based on your before-and-after analysis of a single AI cycle, use the space below to note new insights on whether (and how) this project will lead to scaling AI efforts within your organization.

General Analysis

The Future of AI

Like every other disruptive technology in human history, AI will be a cause for confusion and anxiety, accompanied by actual economic dislocation and hardship for those whose jobs have been made wholly or partially obsolete. At the same time, it will lead to the creation of new jobs and opportunities at an exponential rate. The difference between AI and past disruptions is that these changes will happen far more rapidly—in a few years as opposed to decades or centuries.

Businesses, like individual workers, will experience the same phenomenon. This includes pain and dislocation as AI-fueled automation makes many tasks and processes obsolete. But that is coupled with unprecedented new opportunities for growth and change. Both the pain and the growth are happening simultaneously.

There are many different types of AI,[1] each with the potential to cause simultaneous pain and growth. The AI projects covered in this book (often classed as "Narrow AI") can lead to other, more ambitious ones, but they must do so under human supervision and guidance. Through utilizing machine learning, deep learning,

---

1 Joshi, Naveen. "7 Types of Artificial Intelligence." *Forbes, 19 June 2019,* www.forbes.com/sites/cognitiveworld/2019/06/19/7-types-of-artificial-intelligence/.

and artificial neural networks, today's AI can *simulate* human cognition, often with blinding speed. However, it does not currently have the ability to learn, perceive, and understand like a human being. That ability, known as Artificial General Intelligence (AGI) may someday exist, but today is rudimentary at best. If AGI does emerge in some form, causing both pain and vast potential for growth, the principles for its responsible development and use will remain the same.

Converging Technologies

AI, analytics, and big data are, of course, only one area of transformational technologies that have been described as part of the "Industry 4.0" phenomenon.[2] Others involve ***computerized simulations*** of complex business or manufacturing challenges—and the ***autonomous robotic systems*** designed to meet those challenges. Some of these technologies seem incremental and evolutionary, such as the increased sophistication of ***cloud computing*** and ***system integration,*** while others, like ***the Internet of Things*** (IoT), ***additive manufacturing, augmented reality,*** and the ***"digital human"*** phenomenon,[3] seem much more novel and disrupting. ***Cybersecurity*** is also poised to take on new significance, potentially creating security measures that can outpace malicious actors using AI to cause harm.

Each of these technologies is limited at present, but also have enormous potential to create both disruption and exponential growth. They all are based on the basic principles of artificial intelligence, machine learning, and deep learning. Thus, as their underlying principles evolve and improve, they will increasingly converge.

---

2 Marr, Bernard. "What Is Industry 4.0? Here's a Super Easy Explanation for Anyone." *Forbes, 20 Feb. 2024,* www.forbes.com/sites/bernardmarr/2018/09/02/what-is-industry-4-0-heres-a-super-easy-explanation-for-anyone/.

3 Faber, Tom. "The Rise of Digital Humans." Stage11, 15 Oct. 2021, https://stage11.com/the-rise-of-digital-humans/.

The Quantum Potential

One development that is likely to accelerate the pace of change in all of these areas is known as Quantum AI (QAI), the combination of quantum computing with the science of artificial intelligence.[4] AI and machine learning are dependent on *conventional* computational resources that, despite the persistence of Moore's Law,[5] are inherently limited. Quantum computing has the potential to change that by orders of magnitude. Instead of processing information serially, in bits (ones and zeroes), quantum computers use "qbits," which can represent a one and a zero simultaneously. In theory, such computers could be millions of times faster than today's fastest microchip.

Once it becomes practical, quantum computing will have an unimaginable effect on AI, and on every aspect of converging data and automation technology. This includes vast potential improvements in speed, efficiency, and accuracy. Along with the fear and pain it is likely to generate, QAI and its related will also uncover new and unanticipated business capabilities.

The Key Issue

Whether or not QAI—and the convergence of multiple technologies—will hasten the emergence of "conscious" AGI is debatable. The real issue is the way all such models must be governed. ***Regardless of the underlying computing speed and power,*** AI and all related technology must still be implemented and managed according to the principles described in this book:

- It must be applied according tao the needs of each *type* of organization that uses it.

4 Reichental, Jonathan. "Quantum Artificial Intelligence Is Closer than You Think." *Forbes, 20 Nov. 2023,* www.forbes.com/sites/jonathanreichental/2023/11/20/quantum-artificial-intelligence-is-closer-than-you-think/.

5 Heffernan, Virginia. "Is Moore's Law Really Dead?" *Wired, 22 Nov. 2022,* www.wired.com/story/moores-law-really-dead/.

- It can only be used effectively by organizations that are fundamentally data-ready.
- It will only succeed in the long term if it is conceived and executed sustainably, responsibly, and with regard to the safety and well-being of humans.
- It requires the right people—not only those with technical expertise but also those who can imagine, explore, and manage its true potential.

Above all, no matter how fast or capable AI and related technologies become, they must be implemented with a tangible business and data ***value.*** As powerful tools, they cannot be limited to merely automating routine tasks. The true technology of the future must free humans from trivial, burdensome tasks, and empower them to do more of the creative and innovative decision-making that defines our species at its best. The new technology norm will not come without pain and difficulty, but the benefits are undeniable.

About the Author

Asha Saxena is an adjunct professor at Columbia University and served as an Entrepreneur-in-Residence for the Columbia Business School. She is the founder and CEO of Women Leaders in Data and AI (WLDA), a mastermind networking group for women leaders in the tech world. WLDA offers peer support, industry expert-hosted events, workshops, and guidance on leadership development.

Asha served as CEO of Aculyst Inc., a healthcare analytics firm. Prior to that she served as president and CEO of Future Technologies Inc., (FTI) an international data form specializing in data analytics for Fortune 1000 companies in the US. In 2007, she was invited to be a part of the World Economic Forum, where FTI was named a Global Growth Company.

www.ingramcontent.com/pod-product-compliance
Ingram Content Group UK Ltd.
Pitfield, Milton Keynes, MK11 3LW, UK
UKHW061705190726
13853UKWH00008B/2410

9 798888 456934